ADVANCES IN SILAGE PRODUCTION AND UTILIZATION

Edited by **Thiago da Silva**
and **Edson Mauro Santos**

Advances in Silage Production and Utilization
http://dx.doi.org/10.5772/61574
Edited by Thiago da Silva and Edson Mauro Santos

Contributors

Juliana Oliveira, Thiago Bernardes, João Paulo De Ramos, Gonzalo Ferreira, Alston Brown, Miroslav Hutňan, Carlos Rabelo, Lucas Mari, Ricardo Reis, Jaime Salinas-Chavira, Natthawud Dussadee, Yuwalee Unpaprom, Rameshprabu Ramaraj, Peiqiang Yu

Published by InTech

Janeza Trdine 9, 51000 Rijeka, Croatia

Publishing Process Manager Dajana Pemac

Technical Editor SPi Global

Cover InTech Design team

First published November, 2016

Printed in Croatia

Additional hard copies can be obtained from orders@intechopen.com

Advances in Silage Production and Utilization, Edited by Thiago da Silva and Edson Mauro Santos

p. cm.

Print ISBN 978-953-51-2777-2
Online ISBN 978-953-51-2778-9

For EU product safety concerns:
IN TECH d.o.o., Prolaz Marije Krucifikse Kozulić 3, 51000 Rijeka, Croatia,
info@intechopen.com or visit our website at intechopen.com.

Meet the editors

Dr. Da Silva is a professor of Forage Crops and Pastures at the Federal University of Goias. Dr. Da Silva was a postdoc fellow at the Federal University of Bahia and Federal University of Vicosa where he worked with silage fermentation and ruminant nutrition. Dr. Da Silva has a PhD degree in Animal Science (forage crops and ruminant nutrition) from the Federal University of Vicosa, an MSc degree in Animal Science (forage crops and ruminant nutrition) from the Federal University of Paraiba, and a BS degree in Agronomy Engineering from the State University of Santa Cruz. The fields of research explored by Dr. Da Silva include silage fermentation and microbiology, pasture management, feedlot cattle and sheep, alternative feeds for ruminants, and ruminant nutrition.

Dr. Santos is a professor of Forage Crops and Pastures and Beef Cattle at the Federal University of Paraiba. Dr. Santos received his PhD in Animal Science (forage crops and ruminant nutrition) from the Federal University of Vicosa, a master's degree in Animal Science (forage crops and ruminant nutrition), and a BS degree in Animal Science from the Federal Rural University of Rio de Janeiro. Dr. Santos' areas of interest include silage microbiology, cultivation and conservation of forage crops in semiarid regions, alternative feeds for ruminants, and pasture management.

Contents

Preface

Nowadays, we face a context of large environmental impacts from the livestock systems, and it has changed the perception about just increase of the production. In this new context, we should focus not only on production but also on efficiency of the whole system. The sustainable intensification of livestock systems is a new approach to achieve the efficient use of the resources by reducing feed cost, decreasing competition for food with humans, contributing to decrease nutrient input from nonlocal sources, and contributing to the feed supply.

Forage conservation techniques have been used to intensify the animal production and to increase the efficiency of the whole livestock system as well. Ensiling is an old technique used to store food, mainly vegetable crops, to feed the herd when the forage supply from the pastures is not enough to maintain the productive performance of the ruminant animals. The main principle of silage is anaerobic environment and fermentation of the water-soluble carbohydrates in the fresh crop by the epiphytic lactic acid bacteria (LAB) with the production of lactic acid. However, different fermentation pathways may occur into the silo environment, depending on the availability of substrate, the predominant microbial populations, the dry matter (DM) content, and the buffering capacity of the crop at the ensiling. The main forage crops used for silage production are corn, sorghum, alfalfa, and grasses. Over the years, there have been many efforts on improving silage production and its utilization through studies about fermentation and the silage management to minimize the DM losses and to increase the efficiency of the process.

This book covers the main advances in silage production and utilization, with nine chapters written by internationally recognized experts from different regions of the world with different environmental contexts but with one common objective: to report the most recent findings in their topic. Among four sections, the first one includes three chapters that show the advances in silage management with a deep survey of the use of bacterial inoculants in silages, the influence of the environment on the corn plants for silage production, and the new technologies on silage covering. The second section brings two chapters about the use of nonconventional crops for silage production in tropical areas, which have a significant impact in arid and semiarid regions and contribute to sustainably intensification of livestock systems in those regions. In the third section, two chapters discuss the nutritional aspects: the effects of fermentation products on intake and digestibility of silages and the strategies to maximize fiber utilization of silages in ruminants. The fourth section includes two chapters about the use of silage as a substrate for biogas production in Europe.

The authors of this book have summarized a large amount of research papers and results that provide a consistent explanation of the technical aspects of silage production and utilization. The intended audience are undergraduate and graduate students, scientists, professors, farmers, consultants, and industrial representatives.

We express our appreciation to the authors that made the publication of this book possible. We also recognize that there is much more information that could be discussed, but it is just an introduction of recent advances and new discussions about silage production and its utilization. As editors, we hope that this book can contribute with the knowledge construction and dissemination, as well as the incentivization for the development of more new technologies that could contribute to the efficiency of the livestock systems.

Thiago da Silva
Federal University of Vicosa,
Brasil

Edson Mauro Santos
Federal University of Paraiba,
Brasil

Section 1

Silage Management

Chapter 1

Survey About the Use of Bacterial Inoculants in Brazil: Effects on Silage Quality and Animal Performance

Carlos H.S. Rabelo, Lucas J. Mari and Ricardo A. Reis

Additional information is available at the end of the chapter

http://dx.doi.org/10.5772/64472

Abstract

Our objective was to report the effect of bacterial inoculants on silage quality and animal responses in Brazil. A survey of bacterial inoculants utilization in Brazil was made based on a total of 178 published articles assessing a widely varied crops (alfalfa, cabbage, cassava, corn, grass, high-moisture corn (HMC), high-moisture sorghum, millet, oat, orange bagasse, peanut forage, sorghum, soybean, stylosantes Campo Grande, sugarcane, and sunflower). Sugarcane and grass silages comprised 58.1% of the total crops investigated. Homolactic inoculation reduced dry matter (DM) losses in alfalfa silages, but not in corn, grass, HMC, and sorghum silages. Heterolactic inoculation enhanced the aerobic stability of corn and HMC silages. The use of heterofermentative lactic acid-bacteria (LAB) was more effective to improve fermentation of sugarcane silages compared to homofermentative LAB. Inoculation impaired the DM intake in cattle fed corn, grass, and sugarcane silages, but DM intake increased in sheep due to inoculation. In some cases, silage digestibility was affected by inoculation. Positive responses to inoculation occurred most often when the compatibility between the bacterial inoculant and crop was better understood (e.g., homolactic inoculation for grass silage and heterolactic inoculation for sugarcane silage). The performance of animals consuming inoculated silages has been investigated in Brazil only a few times, but the data suggest a greater impact of bacterial inoculants on DM intake and weight gain in cattle and sheep than that indicated in temperate conditions.

Keywords: aerobic stability, digestibility, fermentation, growth performance, lactic acid bacteria

1. Introduction

Silage is the feedstuff produced by the fermentation of a crop, forage, or agricultural byproduct, usually at greater than 50% moisture content [1]. In Brazil, silage is the principal source of energy and fiber in the diets of dairy cattle [2] and is frequently used in feedlots for the production of beef cattle [3]. However, descriptions of silage production practices and utilization in Brazilian literature are poor [4]. Furthermore, there is a lack of extension programs in Brazil that disseminate and enhance the knowledge of farmers regarding silage management, which has contributed to the production of low-quality products in many cases. As a strategy to alter this scenario, several farmers have chosen to use bacterial inoculants in order to improve silage quality and reduce production costs by decreasing the loss of dry matter (DM). Nevertheless, in Brazil, there are few reviews and surveys concerning the impact of bacterial inoculants on ensiling practices. In addition, the most complete review of this topic (see [5]) indicated that the low number of studies conducted in Brazil at that time did not produce a definitive conclusion about the magnitude of the effect of additives on silage quality and animal performance.

Therefore, our objective was to conduct a survey on the use of bacterial inoculants in Brazil and understand how they affect ensiling processes and animal performance. Here, we highlight that the present survey had an exploratory focus and, because of this, we conducted only a descriptive analysis of the data found in the accessed studies throughout of this text.

2. Bacterial inoculants

Ensiling is the most common method used to preserve a great variety of forages for use during those seasons when the crop is unavailable and/or is decreasing in nutritive value. Ensiling is based on the conversion of simple plant sugars, such as glucose and fructose, to lactic acid by lactic acid bacteria (LAB) under anaerobic conditions [6, 7]. Epiphytic LAB are essential microflora for spontaneous silage fermentation; however, the number and genera of bacteria varies widely in forages [8]. Thus, bacterial inoculants (specifically homofermentative LAB-^{ho}LAB) have been used in order (1) to inhibit the growth of aerobic and undesirable anaerobic microorganisms, (2) promote a rapid decline in the pH of forage after ensiling in order to avoid greater activity of proteases and deaminases derived from its own plant tissues and/or microorganisms, and (3) increase DM recovery [9].

The international literature is rich with data describing the eventual benefits of inoculation. However, no conclusion has been reached about the effect of bacterial inoculants on silage quality and animal performance in Brazil (see [5]) considering previously summarized studies carried out from 1999 to 2009. After 2009, 85 new Brazilian studies (scientific articles published in national and international journals) evaluating the effect of bacterial inoculants for silage production were published (**Figure 1**). Thus, analyzing real life scenarios are important to understand how bacterial inoculants alter silage quality and how they affect the performance of animals consuming inoculated silages.

http://dx.doi.org/10.5772/64472

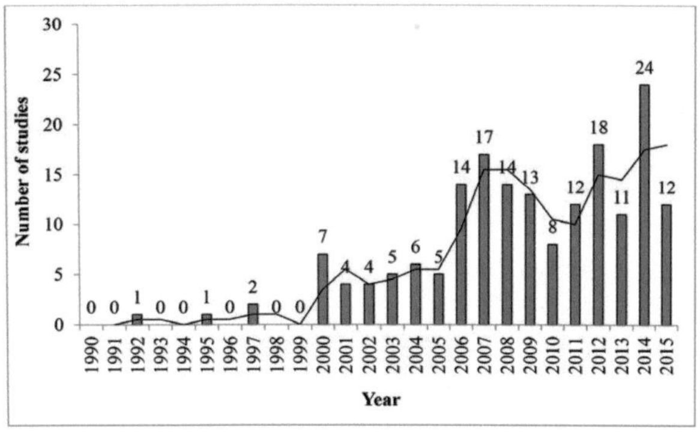

Figure 1. Number of Brazilian articles published concerning bacterial inoculant utilization from the last 26 years (total number of articles accessed = 178).

Initially, the small interest on the topic in the last century in Brazil likely reflected questions about the cost of those inoculants and their effectiveness as in other countries [10], although these are questions that are debated very often. The inconsistent results obtained from early studies carried out in Europe and North America due to low rates of inoculation and questionable viabilities of the bacteria [9], also likely contributed to the initial small interest. Conversely, advances in molecular biology associated with positive responses found across the world may have moved the crescent interest from Brazilian researchers to study bacterial inoculants for silage production. Moreover, the increasing number of techniques used to produce more viable and stable bacteria, and the additional tools developed to access the effects of silage inoculants, may also be part of the reason for the increased interest. Indeed, poor silage management has led to the production of silages of low nutritional value and undesirable sanitary aspects under tropical conditions. Surely, sugarcane and tropical grass silages are still the crops most susceptible to problems that occur during fermentation due to the action of undesirable microorganisms. Thus, these crops comprised 58.1% of all studies evaluated regarding the use of bacterial inoculants (**Figure 2**).

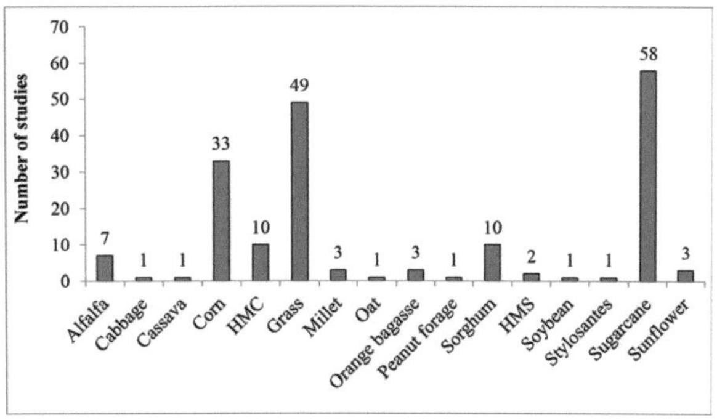

Figure 2. Number of Brazilian studies published regarding the utilization of silage inoculants by crop. *HMC, high-moisture corn; HMS, high-moisture sorghum.

Item	Alfalfa	Corn	Grass	HMC[1]	Sorghum	Sugarcane
One specie						
Bacillus subtilis	–	4	–	–	–	–
Lactobacillus brevis	–	–	–	–	–	27
Lactobacillus buchneri	–	16	8	8	–	62
Lactobacillus hilgardii	–	–	–	–	–	10
Lactobacillus kefiri	–	–	–	–	–	1
Lactobacillus paracasei	–	–	–	–	–	2
Lactobacillus plantarum	–	8	18	9	–	59
Leuconostoc mesenteroides	–	1	–	–	–	–
Streptococcus bovis	–	–	14	–	–	–
Streptococcus faecium	–	–	3	–	–	–
Two species						
L. buchneri + *L. kefiri*	–	–	–	–	–	1
L. buchneri + *Propionibacterium acidipropionici*	–	–	–	1	–	–
Lactobacillus casei + *Streptococcus faecalis*	–	–	–	2	–	–
L. plantarum + *B. subtilis*	–	1	–	–	–	–
L. plantarum + *L. buchneri*	–	4	–	–	–	1
L. plantarum + *Pediococcus acidilactici*	–	6	17	–	–	1
L. plantarum + *Pediococcus pentosaceus*	5	16	3	–	–	7
L. plantarum + *P. acidipropionici*	–	2	2	–	–	4
L. plantarum + *S. faecium*	6	14	12	–	26	3
Combo[2]	4	46	42	16	16	3

[1]HMC, high-moisture corn.
[2]Combination of three or more bacteria.

Table 1. Bacterial species applied in the six main crops used to produce silage in Brazil (number of treatments).

As mentioned earlier, the crescent development in molecular biology techniques has led to a wide range of microbial additives to aid in crop preservation. The LAB (genera *Lactobacillus, Pediococcus, Lactococcus, Enterococcus, Streptococcus,* and *Leuconostoc*) are the main group of bacteria used as silage inoculants, because they all produce lactic acid as a principal product from sugar fermentation [6]. Commonly, the LAB are classified into two groups based on the products of fermenting glucose, as follow: (1) homofermentative (first generation of silage inoculants) → produce two moles of lactic acid from one mole of glucose; and (2) heterofermentative (second generation of silage inoculants) → produce one mole of lactic acid, one mole of carbon dioxide (CO_2), and either one mole of ethanol or one mole of acetic acid from one mole of glucose [11]. However, actually three groups of LAB have been considered [12], as follows: (1) obligate homofermentative → unable to ferment pentoses because the lack enzyme phosphoketolase; (2) facultative heterofermentative → ferment hexoses similarly to the obligate homofermentative but they are able to ferment pentoses; and (3) obligate heterofermentative → ferment hexoses to a range of products. Overall, under most silage conditions where substrate is not lacking, facultative heterofermentative LAB primarily make only lactic acid [9]. Thus, for the sake of simplicity, facultative heterofermentative LAB will be considered part of homofermentative LAB in this review for furthers comparison.

In Brazil, several homofermentative (*Lactobacillus plantarum, L. curvatus, L. acidophilus, L. paracasei, Lactococcus lactis, Leuconostoc mesenteroides, Pediococcus pentosaceus, P. acidilactici, Streptococcus faecium, S. faecalis,* and *S. bovis*) and heterofermentative LAB (*L. buchneri, L. hilgardii, L. kefiri, L. salivarius,* and *L. brevis*) have been used as silage inoculants, leading to different combinations for each crop (**Table 1**). Other microorganisms have also been tested, such as *Propionibacterium acidipropionici, Bacillus subtilis,* and *Saccharomyces* spp., but less frequently.

As described earlier, ^{ho}LAB and heterofermentative LAB (^{he}LAB) comprised first and second generation of silage inoculants, respectively. The ^{ho}LAB gained popularity in the late 1970s and early 1980s because it must quickly grow to dominate silage fermentation reducing DM and nutritive losses [9]. Conversely, homofermentative-inoculated silages often have lower stability during the feed-out phase, because of the greater concentration of lactic acid and residual water-soluble carbohydrates (WSC) [13]. Lactic acid and WSCs are utilized as substrates for the growth of aerobic microorganisms, notably yeasts [13]. Thus, *L. buchneri* was developed as a second generation inoculant to produce acetic acid and improve the aerobic stability of silage by inhibiting the growth of spoilage microorganisms [14]. Nowadays, some commercial silage inoculants contain multiple strains of ^{ho}LAB and often one strain of ^{he}LAB, because of the potential synergistic actions among bacterial strains. For example, previous studies showed that the association between *L. plantarum* and *L. buchneri* accelerated the initial rate of lactic acid fermentation, reducing the pH and causing lower protein degradation, in addition to enhancing the aerobic stability of corn and sorghum silages [13, 15].

In Brazil, ^{ho}LAB were primarily investigated and used as commercial silage inoculants to ensure suitable fermentation (**Figure 3**). Around the year 2000, Brazilian researchers turned their attention and curiosity to investigate the effects of ^{he}LAB on the ensiling of tropical crops, but articles on this topic only started to be published in 2006. Moreover, studies combining ^{ho}LAB and ^{he}LAB started at the same time that second generation silage inoculants were used, but articles evaluating ^{ho}LAB and ^{he}LAB combined started to be published earlier.

Despite the type of silage inoculant used for the six main crops used for ensiling in Brazil, ^{ho}LAB composed the only silage inoculant assessed for alfalfa and sorghum silages (**Figure 4**). Moreover, ^{ho}LAB still composed the majority (>69%) of the treatments for corn, HMC, and grass silages. Sugarcane was the only crop in which ^{he}LAB composed the majority (57%) of the treatments assessed. This scenario is not a surprise, since ^{ho}LAB were primarily investigated and used as commercial silage inoculants in the worldwide, and likely this reflected in a greater number of studies assessing ^{ho}LAB in Brazil. Alfalfa and grass silages often have low WSC content and high buffer capacity, and then pH declines more slowly after the crop is ensiled

[6]. Therefore, is comprehensive why only ^{ho}LAB were assessed for alfalfa and why ^{ho}LAB composed the majority of the treatments for grass. However, considering that corn and sorghum silages that are most susceptible to aerobic deterioration under tropical conditions [16] would be expected a greater number of studies concerning ^{he}LAB or combining ^{ho}LAB and ^{he}LAB to reduce this trouble.

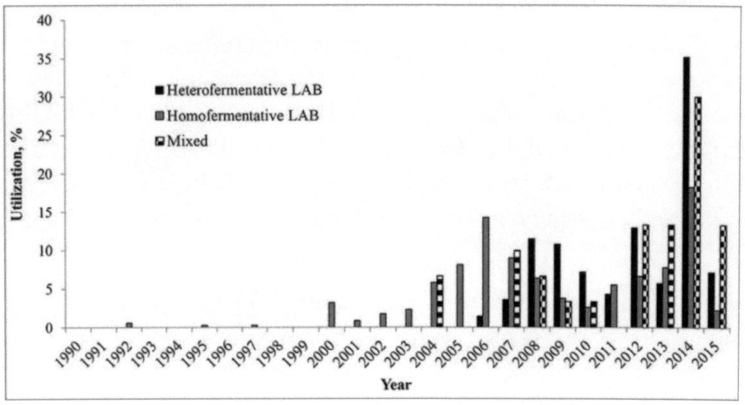

Figure 3. Evolution of the utilization of homofermentative and heterofermentative LAB, either alone or combined (mixed) in Brazil (% related to the number of treatments).

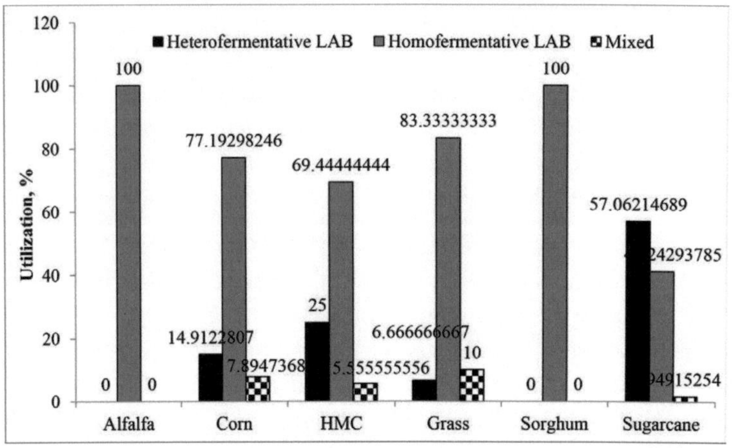

Figure 4. Assessment of homofermentative and heterofermentative LAB, either alone or combined (mixed) by crop in Brazil (% related to the number of treatments containing bacterial inoculants). *HMC, high-moisture corn.

The use of bacterial inoculants has also claimed to improve the nutritive value of silages by reflecting alterations in fermentation patterns, which may be important for tropical silages in particular. The use of tropical forages often results in silages with lower nutritive value than those produced under temperate conditions [16]. Unfavorable aspects of some crops (especially grasses), such as low WSC and DM content (both needed for proper fermentation) at the time of cutting when the highest nutritive value of the grass is achieved and at high buffering capacity, results in poor fermentation and low silage digestibility [17].

Epiphytic LAB utilize carbohydrates as energy and carbon sources for growth, and these microorganisms are only able to convert nonstructural carbohydrates (notably WSCs—mono- and disaccharides) into organic acids, because they lack the enzymatic complex required to metabolize complex polysaccharides [7]. Thus, enzyme-bacterial inoculants may become useful to improve the fermentation patterns and nutritive value mainly of ensiled crops having low WSC content. Bacterial inoculants ensure that LAB will dominate in silage fermentation, whereas the enzymes (i.e., fibrolytic enzymes) contained in those inoculants act on the cell wall, releasing a greater amount of fermentable sugars and increasing substrate availability, thereby improving silage digestibility [18]. Amylolytic and proteolytic enzymes are also commonly used in silage inoculants, and they are particularly useful for cereal silages, reducing the negative effect of the starch-protein matrix on starch digestion in ruminants [19, 20]. Therefore, it is easy to understand why enzyme-bacterial inoculants are used primarily in high-moisture corn (HMC) silages (>55%), followed by grass, corn, sorghum, alfalfa, and sugarcane silages (**Figure 5**). Obviously, the little interest in evaluating enzyme-bacterial inoculants for sugarcane silage is related to the great amount of WSC in this crop, particularly sucrose [21].

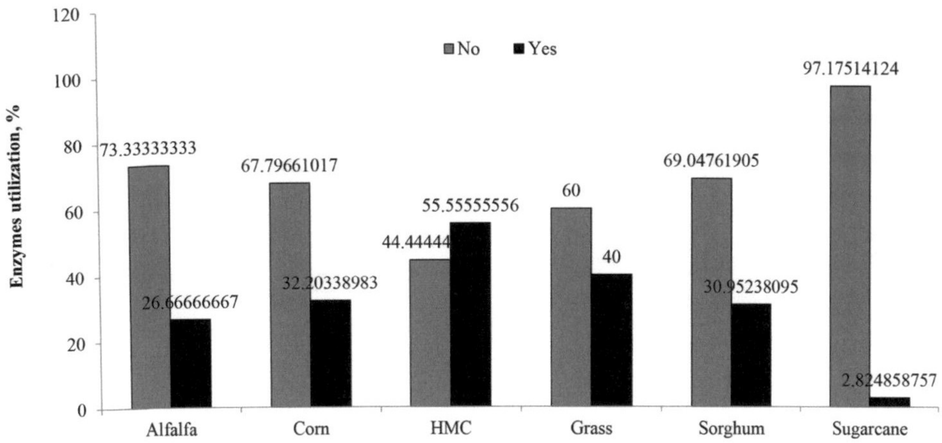

Figure 5. Enzyme utilization in silage inoculants by crop in Brazil (% related to the number of treatments containing bacterial inoculants). *HMC, high-moisture corn.

2.1. Fermentation patterns, nutritive value, and aerobic stability of silages

The use of bacterial inoculants as additives to improve silage fermentation has a long and diverse history. As described earlier, although silage inoculant utilization occurred later in Brazil than Europe and North America, many types and formulations of bacteria are currently sold commercially for this purpose. However, the compatibility between the plant and microorganisms used will determine the success of the application of bacterial inoculants in silages [22]. When that compatibility is better understood, positive responses from inoculation occur more often.

Fermentation and microbiological profile	**CCP**	**Chemical composition**	**CCP**	**Animal performance**	**CCP**
pH	Decreasing	DM, % as fed	Increasing	DMI, kg/day	Increasing
Ammonia-N, % TN	Decreasing	Ash	Decreasing	DMI, % BW	Increasing
WSC	Increasing	EE	Increasing	OMI, kg/day	Increasing
Lactic acid	Increasing	CP	Increasing	NDFI, kg/day	Increasing
Acetic acid	Increasing (heLAB) and decreasing (hoLAB)	NDIN, % N	Decreasing	CPI, kg/day	Increasing
Propionic acid	Increasing	ADIN, % N	Decreasing	Digestible DMI, kg/day	Increasing
Butyric acid	Decreasing	NDF	Decreasing	Digestible OMI, kg/day	Increasing
Total acids[2]	Increasing	ADF	Decreasing	Digestible NDFI, kg/day	Increasing
Lactic:acetic acid	Increasing (hoLAB) and decreasing (heLAB)	Hemicellulose	Decreasing	Digestible CPI, kg/day	Increasing
Ethanol	Decreasing	Cellulose	Decreasing	DM digestibility	Increasing
Total acids:ethanol	Increasing	Lignin	Decreasing	OM digestibility	Increasing
Effluent, kg/t of fresh matter	Decreasing	IVDMD	Increasing	NDF digestibility	Increasing
Gas losses	Decreasing	IVOMD	Increasing	CP digestibility	Increasing
DM losses	Decreasing			Feed efficiency[3]	Decreasing
LAB, log cfu/g of fresh silage	Increasing			ADG, kg/day	Increasing
Yeasts, log cfu/g of fresh silage	Decreasing				
Molds, log cfu/g of fresh silage	Decreasing				
Aerobic stability, h	Increasing				
Maximum temperature, °C	Decreasing				

TN, total nitrogen; WSC, water-soluble carbohydrates; DM, dry matter; LAB, lactic acid bacteria; heLAB, heterofermentative LAB; hoLAB, homofermentative LAB; EE, ether extract; CP, crude protein; NDIN, neutral detergent insoluble N; ADIN, acid detergent insoluble N; NDF, neutral detergent fiber; ADF, acid detergent fiber; IVDMD, *in vitro* DM digestibility; IVOMD, *in vitro* organic matter digestibility; DMI, DM intake; BW, body weight; OMI, organic matter intake; NDFI, NDF intake; CPI, CP intake; ADG, average daily gain.

[1]Adapted from [5].

[2]Total acid content was calculated as the sum of lactic, acetic, and propionic acids.

[3]Feed efficiency was determined by dividing DMI by ADG.

Table 2. Criteria considered as positive (CCP) effect of inoculation for each variable (data are % of DM, unless otherwise stated)[1].

In order to understand the extent to which each type of bacterial inoculant affects silage quality, we summarized data from corn, grass, sugarcane, alfalfa, sorghum, and HMC silages produced in Brazil. All comparisons in this survey were made from studies (at least two studies for each variable) that used a negative treatment (untreated forage—control) against one or more treatments containing bacterial inoculants. Some calculations were made when data were lacking from these publications as follows: hemicellulose content was calculated as neutral detergent fiber (NDF) minus acid detergent fiber (ADF), whereas cellulose content was calculated as ADF minus lignin; the proportion of hemicellulose, cellulose, and lignin were also calculated on a NDF basis; total acid production was calculated as the sum of lactic, acetic, and propionic acids; and the ratio of lactic:acetic acid and total acid:ethanol was also calculated. Butyric acid was not considered in the calculation of total acids because this acid has no beneficial effect on ensiling process [6]. Otherwise, lactic acid (acid more desired to reduce DM loss) and acetic and propionic acids (antifungal properties) have beneficial role during ensiling [6].

As described earlier, we performed only a descriptive analysis of data found in the studies investigated. For that, we did not consider a minimum or maximum time of fermentation to include the data from each study in the final dataset, because our objective was not to show the fermentation pattern regarding the length of ensiling. From the summarized data, the mean, median, standard deviation, and minimum and maximum values were calculated for all variables. Moreover, the frequency of positive responses from inoculation was also calculated, considering only the means declared statistically different in the studies that comprised the database. The difference between the means of untreated and inoculated silages, when there were positive responses, was also calculated. The criteria considered as positive for each variable are given in **Table 2**.

Enterobacteria count was not considered in this survey by lack of data, but it is important to state that enterobacteria are the principal competitors against LAB for sugars after the crop is ensiled, and acetic acid is the principal product of enterobacterial fermentation [8]. Conversely, enterobacteria population often declines after ensiling by influence of anaerobiosis and pH reduction due to the acids produced during fermentation [8].

2.1.1. *Corn silage*

Data were summarized from a total of 29 studies, of which 19, 7, and 7 investigated the effect of ^{ho}LAB, ^{he}LAB, and a combination between both (mixed), respectively. *Bacillus subtilis* was also investigated in two studies. Considering all treatments, the application rate of bacterial inoculants ranged from 5×10^4 to 1×10^9 colony forming units (cfu)/g of fresh forage.

The ranges of fermentation patterns, *in vitro* digestibility, and aerobic stability are given in **Table 3**. Considering the overall mean, lactic acid and silage pH were unaffected by ^{ho}LAB. The concentration of lactic acid was greater by 51.2% when both ^{ho}LAB and ^{he}LAB were applied than observed in untreated silage. The ^{ho}LAB increased by 12.8% the concentration of acid detergent insoluble N (ADIN), suggesting that the temperature of fermentation also increased following inoculation. In addition, ^{ho}LAB slightly reduced (−2.8%) the *in vitro* DM digestibility (IVDMD) of corn silages.

Item[1]	Untreated						Homofermentative LAB						Heterofermentative LAB						Mixed[4]					
	n[2]	Mean	Median	SD[3]	Min	Max	n	Mean	Median	SD	Min	Max	n	Mean	Median	SD	Min	Max	n	Mean	Median	SD	Min	Max
DMoven, % as fed	50	32.99	32.62	3.75	23.07	43.70	67	34.57	34.21	4.16	23.03	48.34	20	32.96	34.05	2.47	25.80	36.71	8	29.73	29.89	3.09	22.95	35.12
Ash	31	4.79	4.20	1.35	3.20	12.68	37	5.27	4.24	1.86	2.71	13.23	8	4.11	4.20	0.28	3.00	4.49	6	3.82	3.83	0.61	3.10	4.52
CP	47	7.22	7.22	0.83	5.29	9.70	64	7.33	7.18	0.75	3.90	10.20	17	7.97	7.50	0.98	6.59	9.43	8	7.43	7.75	0.91	5.50	8.90
NDIN, % N	2	12.64	12.64	12.34	0.30	24.97	2	1.84	1.84	1.37	0.47	3.20	1	-	-	-	-	-	1	-	-	-	-	-
ADIN, % N	4	18.84	20.78	4.61	9.62	24.17	7	21.24	19.55	3.18	17.33	27.00	1	-	-	-	-	-	2	12.89	12.89	3.47	9.42	16.35
NDF	45	53.29	53.38	4.71	38.80	64.93	60	54.48	54.17	5.12	40.00	66.47	17	49.88	49.20	6.08	38.15	62.00	8	51.19	50.11	4.45	39.30	59.50
ADF	25	30.32	29.48	3.75	22.57	39.13	24	31.56	30.20	4.58	23.20	42.52	9	26.60	25.80	2.40	21.80	33.99	8	29.99	30.00	2.99	22.60	36.87
Hemicellulose	25	22.78	22.95	2.78	16.20	31.80	24	24.26	23.86	2.69	17.96	31.37	9	18.63	18.50	1.53	16.30	21.24	8	21.43	19.82	3.21	16.70	32.10
Cellulose	14	25.01	25.30	2.54	17.30	28.93	15	25.23	25.27	2.50	18.50	29.59	4	23.11	22.98	1.74	19.90	26.58	6	24.26	24.98	1.98	19.20	27.04
Lignin	13	6.01	4.90	2.05	2.81	13.05	13	8.38	7.50	3.44	2.53	13.11	4	4.63	4.10	1.39	2.90	7.41	6	4.71	4.54	0.90	2.70	7.41
IVDMD	15	61.17	65.23	7.02	46.40	70.67	19	59.43	65.00	9.75	46.14	71.57	2	69.73	69.73	1.53	68.20	71.26	5	63.96	64.93	5.64	49.87	71.68
IVOMD	6	69.78	68.90	4.62	64.30	78.20	10	71.25	71.35	3.65	64.50	78.20	1	-	-	-	-	-	1	-	-	-	-	-
Effluent, kg/t	7	12.25	10.08	7.21	1.74	28.72	12	12.85	12.86	8.56	1.47	35.43	3	4.83	4.86	0.43	4.19	5.45	0	-	-	-	-	-
Gas losses	3	2.62	2.89	1.72	0.04	4.93	2	0.03	0.03	0.01	0.03	0.04	7	4.18	4.80	1.36	2.27	5.88	0	-	-	-	-	-
DM losses	12	4.74	5.60	2.74	0.55	9.21	13	4.80	5.20	1.27	1.37	6.21	12	6.21	5.53	2.06	2.50	14.01	3	5.25	3.92	3.26	1.70	10.14
1.2-propanediol	3	0.35	0.29	0.27	0.00	0.75	0	-	-	-	-	-	5	0.60	0.46	0.35	0.20	1.26	0	-	-	-	-	-
Lactic acid	13	4.93	4.30	1.69	1.47	7.97	12	4.69	5.08	1.66	1.00	8.01	20	4.90	4.78	1.16	2.17	6.88	5	7.46	7.89	1.64	4.46	9.30
Acetic acid	13	2.46	1.91	1.63	0.33	12.48	12	1.41	1.24	0.77	0.46	2.83	20	2.01	1.18	1.53	0.58	12.57	5	6.66	5.50	4.49	1.50	17.89
Propionic acid	8	0.26	0.05	0.30	0.01	1.46	12	0.23	0.02	0.25	0.01	0.62	13	0.36	0.20	0.31	0.01	1.33	1	-	-	-	-	-
Butyric acid	7	0.06	0.04	0.05	0.00	0.19	12	0.04	0.01	0.05	0.00	0.22	11	0.11	0.07	0.06	0.02	0.22	1	-	-	-	-	-
Total acids[5]	7	6.69	6.65	2.12	3.56	10.15	12	6.33	6.26	1.34	3.58	10.13	11	6.99	7.79	1.26	4.25	8.42	1	-	-	-	-	-
Ethanol	7	0.86	0.70	0.52	0.01	1.84	12	1.62	0.50	1.57	0.01	4.24	9	1.75	1.54	0.72	0.69	2.75	1	-	-	-	-	-
Lactic:acetic acid	13	4.45	3.33	2.97	0.34	14.89	12	5.77	4.26	3.83	0.39	10.90	20	4.07	3.84	1.65	0.34	8.15	5	1.90	1.69	0.79	0.36	2.97
Total acids:ethanol	5	149.73	9.50	227.36	1.93	718.13	11	122.12	12.17	177.89	1.81	674.85	7	4.96	2.16	3.34	1.84	10.81	0	-	-	-	-	-
pH	26	3.84	3.79	0.17	3.42	4.30	28	3.89	3.94	0.20	3.30	4.38	21	3.92	3.91	0.13	3.67	4.16	6	3.88	3.94	0.19	3.53	4.10
Ammonia-N, % TN	22	7.77	4.98	5.18	0.11	30.36	25	6.93	4.43	4.14	1.46	25.72	15	6.39	5.99	2.50	0.09	11.20	4	6.03	5.08	3.12	2.90	11.07
LAB, log cfu/g	9	7.57	8.43	1.33	5.45	9.14	12	8.07	8.41	0.66	6.68	9.28	5	6.85	6.64	0.61	5.88	7.75	0	-	-	-	-	-
Yeasts, log cfu/g	12	5.51	5.25	0.96	4.00	7.67	12	6.02	6.07	1.15	3.00	8.83	10	2.95	2.94	1.25	1.20	4.84	3	4.04	3.79	0.53	3.50	4.83
Molds, log cfu/g	7	3.32	3.70	1.40	1.07	5.69	0	-	-	-	-	-	14	2.95	2.80	1.08	0.00	5.14	2	4.64	4.64	0.44	4.20	5.07
Aerobic stability, h	10	36.49	32.90	21.77	0.30	92.20	3	31.08	33.00	7.89	19.25	41.00	12	109.87	89.95	51.58	26.91	228.00	5	27.65	12.08	23.76	0.20	60.50
Maximum T, °C	7	29.97	27.75	3.51	26.00	42.00	13	29.32	27.75	2.54	25.25	38.00	0	-	-	-	-	-	1	-	-	-	-	-

[1]DM, dry matter; CP, crude protein; NDIN, neutral detergent insoluble N; ADIN, acid detergent insoluble N; NDF, neutral detergent fiber; ADF, acid detergent fiber; IVDMD, *in vitro* DM digestibility; IVOMD, *in vitro* organic matter digestibility; TN, total nitrogen; LAB, lactic acid bacteria.

[2]Number of means.

[3]Standard deviation.

[4]Silages inoculated with both heterofermentative and homofermentative bacteria.

[5]Total acid content was calculated as the sum of lactic, acetic, and propionic acids.

Table 3. Range of fermentation patterns, nutritive value, and aerobic stability of untreated and inoculated corn silages (data are given in % of DM, unless otherwise stated).

All silages were close to or inside the ideal range of the DM content (30–35% of DM) recommended for the production of corn silage [6]. Under these conditions, corn plants often exhibit a great amount of WSC and have a low buffer capacity since well managed. Thus, the lack of positive results from homolactic inoculation is likely related to the desired characteristics of corn plants used at ensiling, once all silages (including the untreated) produced a suitable quantity of lactic acid, with an ideal range between 4 and 7% of the DM [23].

Overall, although positive responses from ^{ho}LAB inoculation were not observed, ^{ho}LAB might be useful to increase lactic acid production and improve fermentation when silage is produced with corn plants harvested with moderately to high DM content (i.e., >37%), because a lack of moisture in dry forages restricts the overall fermentation process [6]. Furthermore, the quality of corn silage produced under tropical conditions is not properly a problem, even though its quality often is lower than that produced under temperate conditions [16]. The main problem of corn silage produced under tropical conditions is related with aerobic deterioration [16] when the silos are opened. The elevated temperature occurring in tropical weather is favorable to yeasts' overgrowth [24], which initiates the spoilage of silages by using residual WSC and lactic acid as substrate to growth, with consequent reduction in the nutritive value of silages. In this regard, ^{he}LAB should be useful to reduce aerobic deterioration of corn silages, but in general, ^{ho}LAB composed 77.2% of all treatments concerning silage inoculants for corn silage. The greater ^{ho}LAB utilization likely still reflects the fact that homolactic inoculants were primarily developed as silage additives, and commercial products based on ^{ho}LAB are most available to be assessed compared with ^{he}LAB.

Despite heterolactic inoculation, acetic acid was unaffected, but the aerobic stability of silages was enhanced (+73.4 h), likely because of reductions in the number of yeasts. Nevertheless, ^{he}LAB increased ethanol production, gas, and DM losses during fermentation by 103.6, 59.7, and 31.2% compared with untreated silages, respectively. Extensive heterolactic fermentation unavoidably increases DM loss during the time the silo is closed, because additional products (i.e., acetic acid, ethanol, and CO_2) are formed besides lactic acid [11]. Furthermore, the concentration of 1,2-propanediol increased 116.7% in silages inoculated with ^{he}LAB. *L. buchneri* comprised the main ^{he}LAB evaluated in corn silage, and this bacterium is able to produce 1,2-propanediol, coupled with acetic acid, during anaerobic degradation of lactic acid [25]. The ammonia-N concentration of silages inoculated with ^{ho}LAB and ^{he}LAB, either alone or combined, is in agreement with well-fermented corn silages (range from 5 to 7%) [23].

Considering the overall means, ^{he}LAB reduced the NDF content of silages by 6.4%. In many cases, the reductions in NDF content have been attributed to the capacity of *L. buchneri* to produce ferulate esterase, an enzyme that acts on cell wall-releasing ferulic acid [27]. However, only some specific strains of *L. buchneri* have the capacity to produce ferulate esterase [26]. Moreover, a net hydrolysis of hemicellulose did not occur when the values were compared on an NDF basis (C.H.S. Rabelo and R.A. Reis). Thus, the reasons for reduced NDF content of corn silages inoculated with ^{he}LAB are still unclear; once DM loss increased, it did not provide better preservation of WSC, which could decrease NDF content by the concentration effect. Heterolactic inoculation also improved IVDMD by 14%, most probably due to a reduction in NDF content.

Although few studies combining ^{ho}LAB and ^{he}LAB were carried out in Brazil, overall means revealed increased lactic and acetic concentration when both inoculants were applied on corn silage compared to untreated silage. Combining ^{ho}LAB and ^{he}LAB may ensure a better fermentation process of corn silage with increased lactic and acetic acid concentration [13], as reported earlier. Consequently, a reduction in DM losses with an increased aerobic stability should be expected, but was not observed. Otherwise, silages treated with both ^{ho}LAB and ^{he}LAB slightly lowered NDF content and increased IVDMD. Even though the data of this survey about combining ^{ho}LAB and ^{he}LAB are not encouraged, most likely due to the low number of studies, further researches should consider the investigation of both ^{ho}LAB and ^{he}LAB for corn silage. The international literature has found a better fermentation process of corn silage accompanied of a greater aerobic stability when hoLAB and heLAB were simultaneously used [13, 14, 15]. These responses may ensure most suitable nutritive value of silage and lead some beneficial on animal response.

The frequency and difference of positive responses found in corn silages from homolactic and heterolactic inoculations are given in **Table 4**. Considering only homolactic inoculation, the greatest frequency of positive responses occurred for aerobic stability, lactic acid, DM content, IVDMD, number of yeasts, and IVOMD. Furthermore, the greatest differences in response were observed for lactic acid, effluent production, and aerobic stability. The greater frequency of positive responses observed for aerobic stability is likely to be related to the low number of trials used to generate the data. According to **Table 3**, the average aerobic stability was greatest for heLAB among all treatments. Conversely, increases in the concentration of lactic acid, DM content, and IVDMD suggest better preservation of soluble sugars during ensiling. The hoLAB have been used to reduce variation in the ensiling process, usually by accelerating the post-ensiling decline in pH, while improving DM and nutrient retention [28].

Item[1]	Homofermentative LAB					Heterofermentative LAB				
	Number of treatments		Mean		Difference, %	Number of treatments		Mean		Difference, %
	Total	Positive responses, %	Untreated	Inoculated		Total	Positive responses, %	Untreated	Inoculated	
DMoven, % as fed	67	38.8	36.54	38.05	+ 4.1	20	5.0	31.70	33.40	+ 5.4
Ash	37	10.8	4.65	3.96	- 14.8	8	0.0	-	-	-
CP	64	17.2	6.58	7.35	+ 11.7	17	5.9	8.30	9.20	+ 10.8
NDIN, % N	2	0.0	-	-	-	1	0.0	-	-	-
ADIN, % N	7	0.0	-	-	-	1	0.0	-	-	-
NDF	60	15.0	57.02	50.68	- 11.1	17	5.9	49.90	43.50	- 12.8
ADF	24	8.3	38.56	35.36	- 8.3	9	0.0	-	-	-
Hemicellulose	24	0.0	-	-	-	9	11.1	23.80	18.80	- 21.0
Cellulose	15	13.3	27.45	24.16	- 12.0	4	0.0	-	-	-
Lignin	13	0.0	-	-	-	4	25.0	4.10	2.90	- 29.3
IVDMD	19	36.8	62.31	66.97	+ 7.5	2	50.0	56.00	68.20	+ 21.8
IVOMD	10	30.0	64.80	76.60	+ 18.2	1	100.0	65.90	71.30	+ 8.2
Effluent, kg/t	12	8.3	28.72	14.25	- 50.4	3	0.0	-	-	-
Gas losses	2	0.0	-	-	-	7	0.0	-	-	-
DM losses	13	30.8	7.95	6.19	- 22.2	12	0.0	-	-	-
WSC	1	0.0	-	-	-	0	0.0	-	-	-
1.2-propanediol	0	0.0	-	-	-	5	60.0	0.15	0.84	+ 479.8
Lactic acid	12	41.7	2.97	5.72	+ 92.3	21	14.3	2.97	4.42	+ 48.6
Acetic acid	12	0.0	-	-	-	21	47.6	1.02	1.40	+ 37.4
Propionic acid	12	0.0	-	-	-	13	0.0	-	-	-
Butyric acid	12	0.0	-	-	-	12	0.0	-	-	-
Ethanol	12	0.0	-	-	-	9	0.0	-	-	-
Lactic: acetic acid	12	0.0	-	-	-	21	4.8	5.00	3.00	- 40.0
pH	28	10.7	3.63	3.44	- 5.1	21	0.0	-	-	-
Ammonia-N, % TN	25	16.0	16.21	13.44	- 17.1	16	0.0	-	-	-
LAB, log cfu/g	12	8.3	6.80	7.41	+ 9.0	5	60.0	5.74	7.25	+ 26.4
Yeasts, log cfu/g	12	33.3	6.10	5.04	- 17.4	10	50.0	4.35	1.70	- 60.9
Molds, log cfu/g	0	0.0	-	-	-	14	7.1	1.07	0.00	- 100.0
Aerobic stability, h	3	66.7	25.00	37.00	+ 48.0	12	75.0	49.86	130.66	+ 162.0

[1]DM, dry matter; CP, crude protein; NDIN, neutral detergent insoluble N; ADIN, acid detergent insoluble N; NDF, neutral detergent fiber; ADF, acid detergent fiber; IVDMD, *in vitro* DM digestibility; IVOMD, *in vitro* organic matter digestibility; TN, total nitrogen; LAB, lactic-acid bacteria.

Table 4. Summary of positive responses of silage inoculants on the fermentation patterns, nutritive value, and aerobic stability of corn silages (data are given in % of DM, unless otherwise stated).

Despite heterolactic inoculation, greater frequencies of positive responses were observed for IVOMD, aerobic stability, number of LAB and yeasts, IVDMD, and acetic acid. In addition, the greatest magnitudes of response were observed for the concentration of 1,2-propanediol, aerobic stability, and molds.

The low number of means for some variables contributed to large values for the frequency of positive responses, as well as the difference between untreated and inoculated silages. However, the data clearly showed that ^{he}LAB in corn silage, composed mainly of *L. buchneri*, were biologically effective. *L. buchneri* has been shown to enhance the aerobic stability of silages by increasing the production of acetic acid, which decreases the growth of spoilage microorganisms [29]. Acetic acid has antifungal characteristics [30], and heterolactic inoculation may be particularly important in silages produced under tropical conditions, as elevated temperatures are favorable for yeast growth [24].

2.1.2. *Tropical grass silage*

Data were summarized from a total of 45 studies, of which 40, 4, and 6 investigated the effect of ^{ho}LAB, ^{he}LAB, and a combination of both (mixed), respectively. In these studies, several tropical grasses were investigated: 18 studies with *Pennisetum purpureum* (Elephant grass cv. Napier and Cameroon), 12 studies with *Panicum maximum* (Guinea grass cv. Mombasa and Tanzania), 11 studies with *Brachiaria brizantha* (Palisadegrass cv. Marandu, Xaraes, and Piata), 3 studies with *Cynodon dactylon* (Bermudagrass), 2 studies with *Cynodon nlemfuensis* (Stargrass), and 1 study with *Brachiaria decumbens*. Considering all treatments, the application rate of silage inoculant ranged from 5×10^4 to 8×10^{10} cfu/g of fresh forage.

The range of fermentation patterns, *in vitro* digestibility, and aerobic stability are given in **Table 5**. Considering the overall mean, homolactic inoculation increased the concentration of lactic acid by 29.4%, leading to a pH drop from 4.75 (untreated silage) to 4.47. The main purpose to use ^{ho}LAB is ensuring a rapid pH decline in earlier times of fermentation (often the first 2 days of ensiling) because the greater production of lactic acid [6]. Indeed, pediococci, streptococci, and lactobacilli comprised the majority commercial homolactic inoculants investigated in Brazilian studies, and they lead to the rapid production of lactic acid and great sugar-to-lactic acid conversion efficiency [6]. Otherwise, after the stable phase of fermentation is reached, similar pH can be reported between untreated and inoculated silage with ^{ho}LAB [6]. The DM losses and ammonia-N concentration decreased 11.4 and 11.7%, respectively, due to the use of ^{ho}LAB. The reduction observed for ammonia-N is likely due to a rapid drop in pH, avoiding proteolysis by the plant, and the action of undesirable microorganisms, such as clostridia. Furthermore, the ADIN content decreased 15.1% due to homolactic inoculation. Results from the present survey agree with the international literature, wherein inoculation with ^{ho}LAB generally results in a faster rate of fermentation, less proteolysis, more lactic acid, less acetic and butyric acids, less ethanol, and a greater recovery of energy and organic matter(OM) [9]. Moreover, the data from this survey suggest that homolactic inoculation is most effective in tropical grass silages, compared to other crops. Homolactic inoculation was also most effective in improving the fermentation process of grass silages, compared with corn and sorghum silages in temperate climates [31]. The reasons for that are because the reduced

WSC concentration and epiphytic bacteria populations found prior to ensiling in those crop, which commits the ensiling process [31]. In our survey, although homolactic inoculation consistently improved the fermentation parameters of tropical grass silages, a small effect was observed on the nutritive characteristics, and IVDMD was only slightly improved (+1.5%).

In some cases, adding homolactic inoculants reduced the aerobic stability of silages, because the lactic acid they produce is used as a growth substrate by yeasts that initiate spoilage [32]. However, unexpectedly the aerobic stability of tropical grass silages increased from 59.5 to 114 h when ^{ho}LAB were applied at ensiling, which is likely to be due to the greater production of acids and a lower pH, inhibiting the growth of aerobic microorganisms. But this is only a hypothesis and perhaps factors other than fermentation end products likely contributed to increase the aerobic stability of grass silages treated with ^{ho}LAB.

Item[1]	Untreated						Homofermentative LAB						Heterofermentative LAB						Mixed[4]					
	n[2]	Mean	Median	SD[3]	Min	Max	n	Mean	Median	SD	Min	Max	n	Mean	Median	SD	Min	Max	n	Mean	Median	SD	Min	Max
DMoven, % as fed	49	23.91	21.50	5.73	14.02	54.90	78	24.19	22.20	5.97	14.13	50.00	3	24.33	27.45	4.22	18.00	27.54	10	23.94	23.17	2.78	18.20	29.70
Ash	17	10.82	11.00	1.80	6.99	16.20	24	10.89	10.75	1.59	6.92	20.60	0	-	-	-	-	-	6	9.53	10.07	0.97	7.10	10.48
CP	55	8.27	7.07	2.78	2.40	43.55	85	8.49	7.74	2.50	2.20	45.85	5	6.16	6.20	1.49	3.90	8.10	10	9.99	11.02	2.81	5.60	16.35
NDIN, % N	4	23.77	24.19	11.27	6.51	40.20	3	21.03	16.01	13.18	6.27	40.80	0	-	-	-	-	-	1	-	-	-	-	-
ADIN, % N	14	17.92	12.71	8.33	5.58	52.49	15	15.21	14.73	4.15	5.27	27.10	0	-	-	-	-	-	1	-	-	-	-	-
NDF	49	69.53	70.20	5.69	47.98	81.97	72	69.22	69.44	4.97	48.59	81.00	5	72.74	73.10	3.51	67.40	79.90	10	77.09	78.96	4.52	64.98	83.20
ADF	44	41.45	44.33	6.60	20.93	52.37	65	41.04	42.48	6.43	21.47	54.40	1	-	-	-	-	-	10	44.41	43.50	3.70	39.64	50.91
Hemicellulose	41	28.62	28.81	4.06	20.97	40.92	63	27.98	29.44	4.79	-0.25	38.88	1	-	-	-	-	-	10	32.68	32.19	3.53	25.34	38.18
Cellulose	21	38.47	38.20	3.80	26.50	45.80	29	38.59	37.65	2.89	32.76	45.70	0	-	-	-	-	-	8	33.03	30.84	5.38	27.10	43.79
Lignin	21	7.24	6.83	1.95	4.42	12.47	29	7.33	7.30	1.71	4.06	11.82	0	-	-	-	-	-	8	10.55	11.38	3.18	4.25	14.85
IVDMD	18	56.23	56.27	6.47	38.30	74.54	27	57.08	60.69	7.63	32.52	67.80	1	-	-	-	-	-	6	69.49	70.98	4.98	56.60	75.54
IVOMD	3	57.87	58.40	1.18	56.10	59.10	3	59.77	58.30	1.96	58.30	62.70	0	-	-	-	-	-	0	-	-	-	-	-
Effluent, kg/t	16	31.74	25.70	18.25	4.90	68.50	20	31.55	24.85	16.75	3.50	61.00	1	-	-	-	-	-	6	40.23	30.60	17.26	26.90	92.00
Gas losses	17	6.06	6.70	3.65	0.28	16.20	16	5.31	4.15	3.18	0.53	14.70	1	-	-	-	-	-	6	0.85	0.54	0.68	0.26	2.90
DM losses	22	12.45	10.10	5.42	2.90	25.38	31	11.03	8.31	5.18	2.10	24.60	1	-	-	-	-	-	2	14.09	14.09	6.09	8.00	20.18
WSC	3	1.37	1.80	0.58	0.50	1.82	3	1.43	1.20	0.77	0.50	2.58	0	-	-	-	-	-	0	-	-	-	-	-
Lactic acid	23	3.87	3.49	1.67	0.05	8.97	35	5.01	4.43	1.70	0.12	10.40	0	-	-	-	-	-	1	-	-	-	-	-
Acetic acid	16	1.36	1.09	0.76	0.30	4.53	28	0.98	0.73	0.59	0.05	3.44	0	-	-	-	-	-	1	-	-	-	-	-
Propionic acid	9	0.38	0.29	0.31	0.00	1.34	17	0.29	0.23	0.20	0.00	1.14	0	-	-	-	-	-	0	-	-	-	-	-
Butyric acid	16	0.06	0.05	0.03	0.00	0.21	24	0.05	0.03	0.03	0.00	0.18	0	-	-	-	-	-	1	-	-	-	-	-
Total acids[5]	7	6.04	6.09	2.19	0.50	10.45	15	7.30	7.25	1.43	0.63	10.49	0	-	-	-	-	-	0	-	-	-	-	-
Ethanol	2	1.17	1.17	0.14	1.04	1.31	4	1.20	0.94	0.41	0.89	2.02	0	-	-	-	-	-	0	-	-	-	-	-
Lactic:acetic acid	16	4.54	3.47	2.90	0.11	10.52	28	15.92	6.63	15.16	0.24	208.00	0	-	-	-	-	-	1	-	-	-	-	-
Total acids:ethanol	2	5.89	5.89	2.63	3.26	8.51	4	8.52	10.10	2.96	2.61	11.26	0	-	-	-	-	-	0	-	-	-	-	-
pH	59	4.75	4.70	0.55	3.36	6.80	80	4.47	4.26	0.54	3.15	6.50	2	4.85	4.85	0.05	4.80	4.90	12	4.51	4.39	0.33	4.00	5.35
Ammonia-N, % TN	52	9.70	8.86	4.28	1.23	39.95	73	8.56	8.50	2.90	0.85	27.78	4	24.59	21.94	15.54	3.20	51.28	9	4.20	3.65	0.98	3.20	8.60
LAB, log cfu/g	5	7.17	7.38	1.04	4.58	8.45	9	8.47	8.57	0.87	7.25	10.03	0	-	-	-	-	-	1	-	-	-	-	-
Yeasts, log cfu/g	4	4.69	4.54	1.68	2.76	6.90	9	3.81	3.70	1.24	2.06	6.70	0	-	-	-	-	-	0	-	-	-	-	-
Aerobic stability, h	2	59.45	59.45	36.55	22.90	96.00	3	114.00	114.00	4.00	108.00	120.00	1	-	-	-	-	-	1	-	-	-	-	-
Maximum T, °C	10	27.53	25.85	3.82	23.00	35.30	10	28.65	29.80	3.86	23.00	40.00	0	-	-	-	-	-	0	-	-	-	-	-

[1]DM, dry matter; CP, crude protein; NDIN, neutral detergent insoluble N; ADIN, acid detergent insoluble N; NDF, neutral detergent fiber; ADF, acid detergent fiber; IVDMD, *in vitro* DM digestibility; IVOMD, *in vitro* organic matter digestibility; WSC, water-soluble carbohydrates; TN, total nitrogen; LAB, lactic-acid bacteria.
[2]Number of means.
[3]Standard deviation.
[4]Silages inoculated with both heterofermentative and homofermentative bacteria.
[5]Total acid content was calculated as the sum of lactic, acetic, and propionic acids.

Table 5. Range of fermentation patterns, nutritive value, and aerobic stability of untreated and inoculated grass silages (data are given in % of DM, unless otherwise stated).

Despite heterolactic inoculation, *L. buchneri* was the only heLAB evaluated in the studies that impaired silage quality by increasing pH, ammonia-N, and NDF and reducing crude protein (CP). The responses to inoculation with *L. buchneri* may be crop specific, as evidenced by a meta-analytical study that showed higher effectiveness when applied in corn silages, compared with grass and small-grain silages [29].

Overall, there were not consistent results by combining hoLAB and heLAB for grass silage. Utilization of both hoLAB and heLAB reduced the pH and ammonia-N concentration in silage; however, DM losses increased by 13.2%. The CP content also increased (+20.8%) following inoculation with both hoLAB and heLAB. Although NDF content increased 10.9% due to inoculation, IVDMD also improved by 23.6%. The number of studies assessing both hoLAB and heLAB as silage inoculants for grass is still very low, but the results reported in this survey suggest a suitable strategy to improve fermentation process along with enhanced silage digestibility.

The ash content of grass silages had an elevated value in all treatments (>9.5%) suggesting contamination, probably by soil, during the ensiling process. Tractors are utilized to transport the harvested forage, fill the silo, and compact the forage mass. Normally, soil in the tractor's tire might be deposited in the forage mass. Moreover, soil contamination is often responsible for the increased number of Clostridia and Bacilli in the ensiling forage [33, 34].

There was no comparison regarding positive responses and differences for heLAB and control silages (**Table 6**), because only a few studies used this group of bacteria (**Table 5**). Homolactic inoculation had the greatest frequency of positive responses for IVOMD, gas losses, acetic acid, lactic acid, and lactic:acetic acid. Furthermore, the greatest differences of response were observed for lactic:acetic acid, yeasts, WSC, and lactic and propionic acids. The increased production of lactic acid allowed by homolactic inoculation reduced gas and DM losses, after CO_2 production ceases and, consequently, preserved a greater amount of soluble sugars, increasing silage digestibility [6].

Regarding association of both hoLAB and heLAB, the greatest frequency of positive responses was observed for butyric acid and DM losses. In addition, the greatest differences in the response observed for the concentration of butyric acid, effluent production, and DM losses is likely to be related to the low number of studies evaluated.

The data from this survey suggest that hoLAB should be the only group used for the ensiling of grass, because this group had the greatest frequency of positive responses compared to heLAB and to utilization of hoLAB and heLAB combined.

2.1.3. Sugarcane silage

Data were summarized from a total of 50 studies, of which 21, 40, and 7 investigated the effect of hoLAB, heLAB, and a combination of both (mixed), respectively. Considering all treatments, the application rate of silage inoculants ranged from 2.5×10^4 to 2.5×10^{10} cfu/g of fresh forage.

The range of fermentation parameters, *in vitro* digestibility, and aerobic stability are given in **Table 7**.

Item[1]	Homofermentative LAB					Mixed[2]				
	Number of treatments		Mean		Difference, %	Number of treatments		Mean		Difference, %
	Total	Positive responses, %	Untreated	Inoculated		Total	Positive responses, %	Untreated	Inoculated	
DMoven, % as fed	78	26.9	22.32	24.55	+10.0	10	20.0	20.90	21.94	+5.0
Ash	24	0.0	–	–	–	6	0.0	–	–	–
CP	85	27.1	7.82	9.36	+19.7	10	10.0	6.16	6.20	+0.7
NDIN, % N	3	0.0	–	–	–	1	0.0	–	–	–
ADIN, % N	15	0.0	–	–	–	1	0.0	–	–	–
NDF	72	9.7	69.99	66.15	−5.5	10	10.0	77.08	64.98	−15.7
ADF	65	9.2	46.23	41.46	−10.3	10	20.0	52.37	50.87	−2.9
Hemicellulose	62	4.8	37.34	35.18	−5.8	10	0.0	–	–	–
Cellulose	29	13.8	38.81	35.40	−8.8	8	0.0	–	–	–
Lignin	29	13.8	5.90	5.20	−11.9	8	0.0	–	–	–
IVDMD	27	18.5	58.17	64.08	+10.2	6	0.0	–	–	–
IVOMD	3	66.7	57.25	60.50	+5.7	0	0.0	–	–	–
Effluent, kg/t	20	0.0	–	–	–	6	16.7	68.50	48.20	−29.6
Gas losses	19	57.9	5.56	3.49	−37.3	6	0.0	–	–	–
DM losses	31	25.8	14.60	9.52	−34.8	2	50.0	10.90	8.00	−26.6
WSC	3	33.3	1.82	2.58	+41.8	0	0.0	–	–	–
Lactic acid	35	48.6	3.59	5.08	+41.7	1	0.0	–	–	–
Acetic acid	28	53.6	1.16	0.74	−36.5	1	0.0	–	–	–
Propionic acid	17	5.9	0.77	1.09	+41.6	0	0.0	–	–	–
Butyric acid	24	37.5	0.05	0.03	−31.2	1	100.0	0.082	0.004	−95.1
Ethanol	4	0.0	–	–	–	0	0.0	–	–	–
Lactic:acetic acid	28	46.4	5.30	10.97	+106.9	1	0.0	–	–	–
pH	80	36.3	4.56	4.20	−8.0	12	0.0	–	–	–
Ammonia-N, % TN	73	30.1	9.97	7.65	−23.3	9	0.0	–	–	–
LAB, log cfu/g	9	44.4	8.36	9.32	+11.4	1	0.0	–	–	–
Yeasts, log cfu/g	9	11.1	5.83	2.06	−64.7	0	0.0	–	–	–
Molds, log cfu/g	4	0.0	–	–	–	0	0.0	–	–	–
Aerobic stability, h	3	33.3	96.00	120.00	+25.0	1	0.0	–	–	–

[1]DM, dry matter; CP, crude protein; NDIN, neutral detergent insoluble N; ADIN, acid detergent insoluble N; NDF, neutral detergent fiber; ADF, acid detergent fiber; IVDMD, *in vitro* DM digestibility; IVOMD, *in vitro* organic matter digestibility; WSC, water-soluble carbohydrates; TN, total nitrogen; LAB, lactic-acid bacteria.
[2]Silages inoculated with both heterofermentative and homofermentative bacteria.

Table 6. Summary of positive responses of silage inoculants on the fermentation patterns, nutritive value, and aerobic stability of grass silages (data are given in % of DM, unless otherwise stated).

Heterolactic inoculants have been used to increase the production of acetic acid in order to reduce aerobic deterioration [28]. For sugarcane silages, the use of ^{he}LAB was proposed to avoid yeast overgrowth and associated ethanol production, with reduced DM losses [35]. Moreover, the reduced DM losses involve a better preservation of WSC [35], which may lead an increased IVDMD of sugarcane silages.

Item[1]	Untreated						Homofermentative LAB						Heterofermentative LAB					
	n[2]	Mean	Median	SD[3]	Min	Max	n	Mean	Median	SD	Min	Max	n	Mean	Median	SD	Min	Max
DMoven, % as fed	52	24.03	23.23	2.96	16.40	33.70	51	23.75	23.96	2.47	15.70	32.60	64	26.10	26.41	2.91	19.00	34.90
Ash	20	4.64	4.74	1.44	1.75	7.36	8	5.24	5.60	1.18	2.80	7.14	20	4.54	4.86	1.46	2.00	7.24
CP	46	3.59	3.55	0.85	1.50	6.70	27	3.83	3.90	0.98	1.74	6.23	43	3.71	3.60	0.72	1.70	7.16
NDIN, % N	2	18.50	18.50	17.01	1.49	35.50	2	11.43	11.43	10.07	1.36	21.50	1	-	-	-	-	-
ADIN, % N	6	10.09	1.57	11.46	1.42	35.97	6	10.93	1.61	12.62	1.32	37.31	5	8.62	1.43	11.61	1.19	37.64
NDF	48	65.63	66.15	5.46	50.19	84.54	39	65.13	64.85	3.06	55.40	75.30	56	62.70	61.95	4.89	48.89	75.90
ADF	41	44.73	43.80	5.49	30.35	58.40	25	41.19	40.36	3.43	34.81	50.38	37	43.00	43.30	5.77	29.21	63.80
Hemicellulose	38	21.02	21.75	3.95	9.80	30.06	22	23.50	23.66	4.00	10.60	30.49	34	20.59	21.35	3.30	5.90	35.40
Cellulose	20	36.11	36.42	4.89	23.93	44.21	13	32.65	31.50	3.31	26.90	39.51	14	35.59	36.85	5.37	22.71	47.40
Lignin	20	8.95	8.05	2.60	4.93	15.20	13	6.87	7.80	1.85	3.21	9.70	14	10.16	8.35	3.46	6.40	16.40
IVDMD	32	48.09	47.30	6.68	32.60	65.40	19	52.24	51.16	4.67	41.10	63.90	30	50.99	48.73	6.65	36.80	69.00
Effluent, kg/t	20	50.86	45.60	27.36	5.40	107.31	3	28.10	29.90	18.34	0.59	53.80	21	48.37	46.10	22.86	2.30	92.18
Gas losses	20	21.14	19.14	6.44	9.43	36.00	2	15.30	15.30	0.50	14.80	15.80	23	18.34	15.91	6.95	8.50	36.20
DM losses	28	24.63	21.91	9.16	6.08	66.00	30	23.73	25.21	5.80	7.69	37.89	35	20.27	18.10	7.27	5.19	66.60
WSC	21	6.65	4.50	4.82	0.74	25.30	47	7.06	3.16	6.34	0.94	33.40	53	7.09	3.19	6.53	1.20	30.80
Lactic acid	27	3.41	3.63	1.43	0.02	6.07	41	4.37	4.50	0.85	0.34	6.63	55	3.09	3.14	0.76	0.07	8.00
Acetic acid	30	2.81	2.39	1.42	0.28	6.75	52	1.53	0.92	1.17	0.20	6.97	57	2.73	2.25	1.47	0.50	8.51
Propionic acid	26	0.46	0.29	0.39	0.00	2.47	48	0.49	0.34	0.30	0.01	1.90	55	0.52	0.40	0.39	0.00	3.91
Butyric acid	20	0.05	0.03	0.05	0.00	0.28	39	0.26	0.19	0.20	0.00	0.99	47	0.22	0.13	0.17	0.00	1.21
Total acids[4]	21	6.34	5.94	2.41	0.62	12.75	37	5.77	5.57	1.15	0.63	11.70	53	6.17	5.24	1.66	3.42	15.57
Ethanol	31	8.94	8.27	4.76	0.44	26.52	53	13.91	15.08	5.08	0.86	21.80	58	6.36	6.29	3.36	0.29	20.62
Lactic:acetic acid	26	1.92	1.25	1.38	0.01	9.00	41	8.00	6.67	5.05	0.36	20.70	54	1.83	1.43	1.07	0.02	5.08
Total acids:ethanol	18	0.84	0.49	0.54	0.23	2.90	35	0.80	0.34	0.76	0.27	6.15	51	1.70	0.78	1.36	0.42	13.76
pH	53	3.57	3.58	0.18	2.94	4.30	65	3.59	3.61	0.15	3.05	4.10	77	3.58	3.61	0.14	2.85	3.90
Ammonia-N, % TN	19	6.55	4.75	3.81	0.65	14.81	24	7.09	6.65	2.61	1.75	14.68	12	6.41	5.90	3.77	0.94	11.95
LAB, log cfu/g	16	7.00	8.01	1.62	2.05	8.90	31	6.75	6.82	0.55	5.65	8.16	49	7.25	7.03	1.37	2.09	9.95
Yeasts, log cfu/g	19	5.53	5.76	0.86	2.47	7.20	34	5.85	6.10	0.48	4.66	6.71	51	4.61	4.81	1.00	2.00	7.20
Aerobic stability, h	12	40.60	33.35	20.51	5.63	97.60	18	29.24	21.43	12.82	16.00	107.00	24	39.74	23.00	24.73	7.63	211.00
Maximum T, °C	9	38.74	39.30	5.06	29.80	47.00	16	42.87	43.50	2.22	31.70	47.20	20	41.53	43.65	3.97	22.50	45.70

[1]DM, dry matter; CP, crude protein; NDIN, neutral detergent insoluble N; ADIN, acid detergent insoluble N; NDF, neutral detergent fiber; ADF, acid detergent fiber; IVDMD, *in vitro* DM digestibility; WSC, water-soluble carbohydrates; TN, total nitrogen; LAB, lactic-acid bacteria.
[2]Number of means.
[3]Standard deviation.
[4]Total acid content was calculated as the sum of lactic, acetic, and propionic acids.

Table 7. Range of fermentation patterns, nutritive value, and aerobic stability of untreated and inoculated sugarcane silages (data are given in % of DM, unless otherwise stated).

In this regard, considering the overall mean, acetic acid was unaffected, but ^{he}LAB reduced the ethanol concentration by 28.9%, because the number of yeasts was reduced. Reductions in yeast growth were probably due to the slight drop in the lactic:acetic acid ratio, in addition to a 12.8% increase in the production of propionic acid, which also has antifungal properties [30]. *L. brevis*, *L. buchneri*, and *L. hilgardii* are the most common ^{he}LAB used in sugarcane silage by Brazilian studies, and they are capable in producing 1,2-propanediol anaerobically [36]. Thus, the greatest production of propionic acid is likely to be related to the conversion of 1,2-propanediol to equimolar portions of 1-propanol and propionic acid, a process driven by

Lactobacillus diolivorans, assuming that this bacterium was present in ensiled forage [37, 38]. Moreover, heterolactic inoculation reduced gas and DM losses by 13.2% and 17.7%, respectively. Fermentative losses decreased because of the control of yeast growth. For each mole of glucose consumed, yeasts produce two moles of ethanol and CO_2, leading to 49% of DM losses in the ethanolic pathway [6]. In addition, *L. buchneri* was the main bacterium used in sugarcane silage, and this bacterium is known for its lack of acetaldehyde dehydrogenase [39], which reduces ethanol production. Conversely, the enhanced aerobic stability caused by heterolactic inoculation did not occur based on the overall mean.

The ADIN content decreased 14.6% due to heterolactic inoculation, suggesting that the control of yeast activity reduced the temperature of the ensiled mass during fermentation. Despite the effects on the fiber fraction, ^{he}LAB reduced the NDF content by 4.5%, likely due to increased hydrolysis of hemicellulose during fermentation [6]. Indeed, a net disappearance of hemicellulose was observed in ^{he}LAB-treated sugarcane silages (**Figure 6**), and as a consequence, the IVDMD increased by 6% on average.

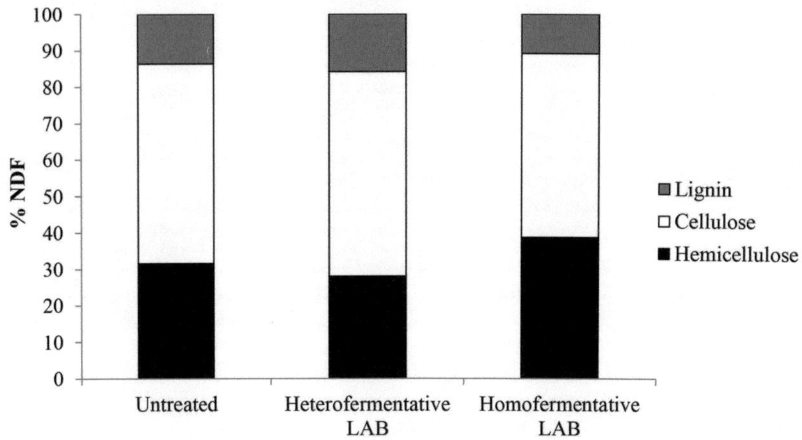

Figure 6. Proportion of hemicellulose, cellulose, and lignin in sugarcane silages untreated or inoculated with homofermentative and heterofermentative LAB (as-is a NDF basis).

The main action of homolactic inoculation is related to the increased preservation of nutrients during fermentation via the production of lactic acid [6]. In this regard, lactic acid increased by 28.2% in sugarcane silages inoculated with ^{ho}LAB. In addition, there was greater preservation of residual WSC (+6.2%), a reduction in the concentration of acetic acid (−45.7%), and a decrease in DM losses (−3.6%). As a consequence, IVDMD improved by 8.6%. However, homolactic inoculation increased ethanol production by 55.5% once yeasts are able to use WSC to grow in anaerobic conditions [8]. Furthermore, homolactic inoculation reduced the aerobic stability of silages by 11.4 h.

The frequency and magnitude of positive responses found in sugarcane silages from homolactic and heterolactic inoculations are given in **Table 8**.

Item[1]	Homofermentative LAB					Heterofermentative LAB				
	Number of treatments		Mean		Difference, %	Number of treatments		Mean		Difference, %
	Total	Positive responses, %	Untreated	Inoculated		Total	Positive responses, %	Untreated	Inoculated	
DMoven, % as fed	51	7.8	24.07	27.50	+ 14.3	63	42.9	24.43	27.25	+ 11.5
Ash	8	12.5	5.24	2.98	- 43.1	20	5.0	6.97	5.91	- 15.2
CP	27	3.7	1.50	2.50	+ 66.7	42	7.1	2.96	3.52	+ 18.9
ADIN, % N	6	0.0	-	-	-	5	0.0	-	-	-
NDF	39	5.1	67.15	60.20	- 10.4	55	12.7	68.50	63.42	- 7.4
ADF	25	8.0	43.56	37.59	- 13.7	36	11.1	45.28	41.25	- 8.9
Hemicellulose	22	0.0	-	-	-	33	0.0	-	-	-
Cellulose	13	7.7	35.87	30.51	- 14.9	14	0.0	-	-	-
Lignin	13	0.0	-	-	-	14	0.0	-	-	-
IVDMD	19	10.5	43.26	52.63	+ 21.7	29	6.9	44.94	54.80	+ 21.9
IVOMD	2	0.0	-	-	-	0	0.0	-	-	-
Effluent, kg/t	3	33.3	6.98	0.59	- 91.5	21	9.5	98.91	78.38	- 20.8
Gas losses	2	0.0	-	-	-	23	30.4	17.31	11.28	- 34.8
DM losses	30	10.0	25.87	25.34	- 2.1	39	48.7	25.66	17.08	- 33.4
WSC	48	8.3	1.31	2.74	+ 109.6	55	25.5	4.28	5.87	+ 37.1
Lactic acid	41	19.5	2.13	3.71	+ 74.5	54	7.4	2.13	2.84	+ 33.6
Acetic acid	52	11.5	1.68	0.59	- 64.7	56	51.8	2.32	3.02	+ 30.1
Propionic acid	48	31.3	0.35	0.55	+ 58.3	54	37.0	0.33	0.65	+ 94.6
Butyric acid	39	2.6	0.07	0.03	- 61.5	46	8.7	0.12	0.03	- 75.5
Ethanol	53	9.4	9.09	2.43	- 73.3	57	56.1	10.39	5.72	- 45.0
Lactic:acetic acid	41	2.4	1.19	1.76	+ 48.4	53	1.9	1.19	0.72	- 39.0
pH	65	9.2	3.53	3.37	- 4.4	76	10.5	3.63	3.43	- 5.7
Ammonia-N, % TN	24	8.3	7.28	7.25	- 0.4	11	27.3	7.28	5.76	- 20.9
LAB, log cfu/g	31	0.0	-	-	-	49	14.3	7.83	9.39	+ 20.0
Yeasts, log cfu/g	34	5.9	6.21	4.23	- 31.9	51	19.6	5.87	3.34	- 43.1
Aerobic stability, h	18	0.0	-	-	-	23	8.7	76.80	151.50	+ 97.3

[1]DM, dry matter; CP, crude protein; NDIN, neutral detergent insoluble N; ADIN, acid detergent insoluble N; NDF, neutral detergent fiber; ADF, acid detergent fiber; IVDMD, *in vitro* DM digestibility; WSC, water-soluble carbohydrates; TN, total nitrogen; LAB, lactic-acid bacteria.

Table 8. Summary of positive responses of silage inoculants on the fermentation patterns, nutritive value, and aerobic stability of sugarcane silages (data are given in % of DM, unless otherwise stated).

Homolactic inoculation had the greatest frequency of positive responses for effluent production and propionic acid, but there is no clear explanation for these results. Furthermore, the greatest difference of responses was observed for WSC, effluent production, and lactic acid. Commercial homolactic inoculants investigated in Brazilian studies were often composed of pediococci, streptococci, and lactobacilli. Thus, the inoculation of silages with pediococci and streptococci leads to the rapid production of lactic acid and great sugar-to-lactic acid conversion efficiency [6, 40]. Afterward, the more acid-tolerant lactobacilli continue producing lactic acid until stable fermentation is achieved [6]. Therefore, the greater production of lactic acid and preservation of WSC from homolactic inoculation in sugarcane silages is expected.

Regarding heterolactic inoculation, the greatest frequency of positive responses was observed for ethanol, acetic acid, and DM losses. In addition, the greatest differences in responses were observed for aerobic stability and propionic acid. Second generation bacterial inoculants are expected to improve the aerobic stability of silages. As described earlier, the bacteria that composed the ^{he}LAB group used for sugarcane ensiling are able to convert lactic acid into acetic acid and 1,2-propanediol [25, 36, 41] when the primary fermentation is ended up. In turn, acetic acid has an antagonistic effect on the growth of yeasts [30], and reductions in ethanol production are expected.

Item[1]	Untreated						Homofermentative LAB					
	n[2]	Mean	Median	SD[3]	Min	Max	n	Mean	Median	SD	Min	Max
DMoven, % as fed	7	41.26	51.29	14.46	14.64	56.20	11	33.39	23.49	15.60	14.81	62.64
Ash	4	11.51	12.47	1.99	7.60	13.49	4	10.54	11.64	1.76	7.02	11.85
CP	7	19.75	19.51	2.00	16.38	24.33	11	20.14	20.49	1.83	15.90	23.44
NDIN, % N	2	13.03	13.03	1.71	11.32	14.73	3	13.47	12.28	1.82	11.93	16.21
ADIN, % N	3	15.04	15.92	2.31	11.57	17.63	7	16.68	17.17	1.55	11.24	19.08
NDF	8	45.06	45.82	3.19	40.18	52.04	13	44.26	43.43	4.18	37.86	54.28
ADF	7	38.29	39.76	2.26	33.99	40.39	9	38.00	39.94	3.00	33.22	42.50
Hemicellulose	6	7.59	6.88	1.99	5.43	13.57	7	8.06	7.25	2.50	4.14	11.78
Cellulose	3	26.42	25.41	1.47	25.22	28.63	5	26.44	25.60	1.53	24.38	29.72
Lignin	4	12.20	11.51	2.16	9.25	16.52	9	13.29	12.71	3.12	8.84	18.87
IVDMD	3	68.92	66.50	5.92	62.46	77.81	7	67.98	65.13	6.22	60.21	75.57
DM losses	2	10.58	10.58	1.09	9.49	11.67	6	5.17	4.95	3.13	1.33	9.55
WSC	3	2.78	2.44	1.00	1.62	4.27	7	3.17	2.97	1.32	1.57	4.84
Lactic acid	3	4.92	4.45	2.82	1.16	9.15	7	7.17	5.62	4.00	0.95	13.83
Acetic acid	3	5.03	3.90	3.51	0.89	10.29	7	5.24	3.93	2.05	2.35	8.36
Propionic acid	3	0.14	0.14	0.10	0.00	0.29	7	0.20	0.10	0.15	0.00	0.41
Butyric acid	3	0.33	0.01	0.43	0.00	0.99	7	1.00	0.02	1.13	0.00	2.85
Total acids[4]	3	10.09	11.74	3.16	5.34	13.2	7	12.61	13.23	2.86	7.97	17.84
Ethanol	3	0.37	0.46	0.23	0.02	0.61	7	1.44	0.51	1.26	0.02	3.08
Lactic:acetic acid	3	2.46	2.40	1.61	0.11	4.87	7	2.26	3.02	1.59	0.12	4.57
Total acids:ethanol	3	89.78	25.29	88.5	21.5	223	7	79.97	32.91	103.67	3.20	442.83
pH	6	4.83	4.66	0.39	4.25	5.50	10	4.98	4.78	0.60	4.22	6.11
Ammonia-N, % TN	6	13.85	8.21	9.50	5.21	29.48	10	22.85	28.61	10.99	5.30	37.27
Maximum T, °C	3	26.85	27.00	1.70	24.30	29.25	7	27.21	27.33	1.18	23.78	28.63

[1]DM, dry matter; CP, crude protein; NDIN, neutral detergent insoluble N; ADIN, acid detergent insoluble N; NDF, neutral detergent fiber; ADF, acid detergent fiber; IVDMD, *in vitro* DM digestibility; WSC, water-soluble carbohydrates; TN, total nitrogen.
[2]Number of means.
[3]Standard deviation.
[4]Total acid content was calculated as the sum of lactic, acetic, and propionic acids.

Table 9. Range of fermentation patterns, nutritive value, and aerobic stability of untreated and inoculated alfalfa silages (data are given in % of DM, unless otherwise stated).

2.1.4. *Alfalfa, sorghum, and high-moisture corn silages*

Data on alfalfa, sorghum, and HMC silages were summarized from 7, 10, and 10 studies, respectively. All studies comprising alfalfa and sorghum evaluated ^{ho}LAB only. For HMC silages, ^{ho}LAB, ^{he}LAB, and a combination between both (mixed) were investigated in six, three, and one study, respectively. Considering all treatments, the application rate of silage inoculant for alfalfa, sorghum, and HMC ranged from 1×10^5 to 9.9×10^5 cfu/g, 9.99×10^4 to 8×10^5 cfu/g, and 5×10^4 to 1×10^6 cfu/g of fresh forage, respectively.

Item[1]	Untreated						Homofermentative LAB					
	n[2]	Mean	Median	SD[3]	Min	Max	n	Mean	Median	SD	Min	Max
DMoven, % as fed	25	30.58	30.89	4.48	19.80	42.33	35	28.44	26.31	4.17	21.70	42.29
Ash	5	5.25	4.20	1.55	3.79	8.77	5	4.88	3.76	1.60	3.27	8.53
CP	18	7.03	7.04	1.40	5.15	13.28	22	7.80	7.55	1.85	5.32	14.08
NDF	23	52.97	53.13	8.87	36.67	73.89	31	56.13	58.67	7.37	35.36	71.42
ADF	19	28.20	23.70	6.78	18.99	44.95	23	31.33	28.77	7.49	19.60	45.78
Hemicellulose	19	22.84	23.20	3.13	14.76	31.65	23	22.90	22.61	2.56	11.68	27.70
Cellulose	14	23.25	21.22	4.51	17.03	39.61	16	24.63	23.37	3.80	17.28	39.99
Lignin	14	3.93	3.42	1.47	1.96	8.34	16	4.58	3.84	2.03	1.99	9.32
IVDMD	16	58.39	59.02	2.54	46.38	62.88	22	59.46	59.77	1.51	55.00	61.75
DM losses	11	1.88	1.69	0.77	0.00	5.12	13	4.18	2.48	2.92	0.31	14.14
WSC	12	1.12	0.32	1.28	0.12	7.34	14	1.49	0.23	2.02	0.14	6.62
Lactic acid	8	5.69	5.20	1.12	3.95	8.54	10	5.80	6.06	1.32	3.90	7.65
Acetic acid	5	1.55	1.52	0.42	0.86	2.42	7	1.53	1.21	0.66	0.82	2.93
Lactic:acetic acid	5	4.38	3.82	1.17	2.89	7.14	7	5.35	3.79	2.50	2.50	8.33
pH	16	3.94	3.86	0.18	3.74	4.94	20	3.94	3.87	0.16	3.66	4.88
Ammonia-N, % TN	15	6.01	4.62	3.76	0.26	16.87	17	5.48	4.05	3.18	0.38	16.79

[1]DM, dry matter; CP, crude protein; NDF, neutral detergent fiber; ADF, acid detergent fiber; IVDMD, *in vitro* DM digestibility; WSC, water-soluble carbohydrates; TN, total nitrogen.
[2]Number of means.
[3]Standard deviation.

Table 10. Range of fermentation patterns, nutritive value, and aerobic stability of untreated and inoculated sorghum silages (data are given in % of DM, unless otherwise stated).

The range of fermentation parameters, *in vitro* digestibility, and aerobic stability in alfalfa, sorghum, and HMC silages are given in **Tables 9, 10,** and **11**, respectively. Considering the overall mean, there was a large difference in the DM content of alfalfa silages, with 33.4% in inoculated silage and 41.3% in untreated silage. Homolactic inoculation increased the concentration of lactic acid by 45.8% in alfalfa silage; however, the pH of silage did not decline, compared with untreated silage; this point may be a consequence of the greater moisture content found in ^{ho}LAB-inoculated silages.

The ^{ho}LAB reduced DM losses by 51.1% in alfalfa silages. Conversely, homolactic inoculation increased the concentration of ammonia-N by 65%, and an increase from 0.37 to 1.44% in the ethanol concentration was also observed. The greater concentration of ammonia-N was unexpected, since lactic acid produced by ^{ho}LAB should be able to decrease proteolytic bacterial populations within the ensiled mass.

Considering the frequency of positive responses of inoculation, only the acetic acid concentration was affected, which was reduced by ^{ho}LAB in 14.3% (−35.95%) of the treatments. Usually, improvements on quality of alfalfa silages have been reported due to the homolactic inoculation [42, 43] most likely due to increases on the numbers of LAB, which is quite low in alfalfa [44]. Although the present survey does not contain data regarding number of LAB in alfalfa silages, homolactic inoculation improved the preservation of this crop.

Item[1]	Untreated						Homofermentative LAB						Heterofermentative LAB					
	n[2]	Mean	Median	SD[3]	Min	Max	n	Mean	Median	SD	Min	Max	n	Mean	Median	SD	Min	Max
DMoven, % as fed	17	64.36	64.43	2.55	52.47	69.22	17	64.86	64.23	3.95	51.52	72.51	8	65.94	65.98	0.50	64.92	66.66
Ash	14	1.42	1.33	0.17	1.27	2.43	13	1.32	1.32	0.06	1.23	1.42	1	-	-	-	-	-
CP	23	9.12	9.23	1.88	6.39	17.32	23	9.02	9.35	1.68	6.62	11.71	5	12.03	10.69	2.51	10.17	18.32
EE	12	3.76	3.54	0.52	2.93	5.21	15	3.94	3.79	0.74	2.71	5.68	0	-	-	-	-	-
ADIN, % N	8	0.00	0.00	0.00	0.00	0.01	8	0.01	0.01	0.00	0.00	0.02	0	-	-	-	-	-
NDF	13	15.84	14.71	6.20	5.53	39.50	15	16.00	10.84	7.55	7.38	32.67	5	6.96	6.24	1.38	5.79	10.42
ADF	17	4.38	4.44	1.69	1.19	7.60	16	4.76	4.85	1.85	0.54	8.32	5	2.19	1.43	1.33	1.04	5.51
Hemicellulose	8	15.54	13.50	5.66	4.34	36.12	8	17.29	17.98	6.81	6.20	29.10	4	4.74	4.77	0.21	4.45	4.97
Gas losses	2	1.93	1.93	0.08	1.85	2.01	0	-	-	-	-	-	8	2.15	2.30	0.70	1.21	2.99
DM losses	9	1.78	1.50	0.87	0.46	3.69	6	1.41	1.02	1.12	0.11	3.29	4	1.54	1.57	0.12	1.33	1.71
Lactic acid	7	1.73	1.36	0.62	1.22	3.90	6	1.27	1.25	0.11	1.13	1.49	4	3.58	3.66	0.21	3.16	3.84
Acetic acid	7	0.19	0.16	0.05	0.15	0.36	6	0.19	0.19	0.02	0.15	0.23	4	0.37	0.36	0.03	0.34	0.42
Propionic acid	7	0.10	0.10	0.02	0.01	0.14	6	0.13	0.13	0.02	0.10	0.15	4	0.02	0.02	0.01	0.01	0.03
Total acids[4]	6	1.65	1.65	0.08	1.47	1.77	6	1.58	1.56	0.09	1.44	1.84	0	-	-	-	-	-
Lactic: acetic acid	6	8.50	8.55	0.91	6.25	9.86	6	6.95	6.85	1.00	5.22	8.94	0	-	-	-	-	-
pH	12	4.02	3.98	0.11	3.83	4.40	8	4.12	4.08	0.19	3.81	4.42	9	4.03	4.02	0.10	3.90	4.20
Ammonia-N, % TN	3	0.56	0.72	0.21	0.24	0.72	0	-	-	-	-	-	4	0.28	0.28	0.02	0.26	0.32
LAB, log cfu/g	2	5.53	5.53	0.62	4.91	6.14	4	6.09	6.44	0.77	4.56	6.92	0	-	-	-	-	-
Yeasts, log cfu/g	4	4.63	4.32	1.25	3.20	6.70	4	5.27	5.76	0.97	3.33	6.23	8	4.30	5.10	1.80	1.34	6.39
Molds, log cfu/g	2	3.67	3.67	0.02	3.65	3.68	0	-	-	-	-	-	8	3.06	2.99	0.67	1.71	4.42
Aerobic stability, h	8	51.91	54.75	10.06	36.00	68.00	6	45.00	45.00	12.50	25.50	60.00	8	154.39	111.15	70.41	86.70	265.00
Maximum T, °C	6	32.80	34.00	3.23	25.20	37.20	6	34.57	37.35	5.64	25.00	40.50	0	-	-	-	-	-

[1]DM, dry matter; CP, crude protein; EE, ether extract; ADIN, acid detergent insoluble N; NDF, neutral detergent fiber; ADF, acid detergent fiber; TN, total nitrogen; LAB, lactic-acid bacteria.

[2]Number of means.

[3]Standard deviation.

[4]Total acid content was calculated as the sum of lactic, acetic, and propionic acids.

Table 11. Range of fermentation patterns, nutritive value, and aerobic stability of untreated and inoculated high-moisture corn silages (data are given in % of DM, unless otherwise stated).

Sorghum silages had few alterations on fermentation parameters due to homolactic inoculation. However, DM losses increased from 1.88 to 4.18% when silages were inoculated, when compared with losses in untreated silage.

The inoculation of sorghum silages also increased the NDF content by 6%, but the ammonia-N concentration decreased by 8.8%. Positive responses from inoculation in sorghum silages occurred only for DM (+14.5%), CP (+15.2%), NDF (−8.5%), and IVDMD (+20.8%) at frequencies of 8.6, 4.6, 6.5, and 9.1%, respectively. Overall, the lack of positive results from inoculation is likely related to the suitable characteristics of sorghum for the ensiling process [45]. Similar to corn, sorghum plants also have good fermentation capability, considerable WSC and DM contents, and low buffer capacity. However, sorghum silages often have low aerobic stability because the suitable characteristics described earlier [13, 15]. Although aerobic deterioration can become a great problem under tropical conditions, there is not any study that assessed heLAB for sorghum silage in Brazil.

Item[1]	Homofermentative LAB					Heterofermentative LAB				
	Number of treatments		Mean		Difference, %	Number of treatments		Mean		Difference, %
	Total	Positive responses, %	Untreated	Inoculated		Total	Positive responses, %	Untreated	Inoculated	
DMoven, % as fed	17	11.8	69.07	71.71	+ 3.8	8	0.0	-	-	-
Ash	13	15.4	1.33	1.24	- 6.4	1	100.0	2.43	2.31	- 4.9
CP	23	8.7	6.50	7.64	+ 17.6	5	0.0	-	-	-
EE	15	6.7	3.16	3.84	+ 21.5	0	0.0	-	-	-
ADIN	8	12.5	0.008	0.005	- 37.5	0	0.0	-	-	-
NDF	15	33.3	21.89	19.70	- 10.0	5	0.0	-	-	-
ADF	16	18.8	2.42	1.23	- 49.2	5	0.0	-	-	-
Hemicellulose	8	37.5	13.41	8.25	- 38.5	4	0.0	-	-	-
Effluent, kg/t	0	0.0	-	-	-	4	0.0	-	-	-
Gas losses	0	0.0	-	-	-	8	0.0	-	-	-
DM losses	6	50.0	2.74	0.31	- 88.6	4	0.0	-	-	-
Lactic acid	6	0.0	-	-	-	4	0.0	-	-	-
Acetic acid	6	0.0	-	-	-	4	0.0	-	-	-
Propionic acid	6	83.3	0.11	0.13	+ 18.3	4	0.0	-	-	-
Lactic: acetic acid	6	16.7	6.25	8.94	+ 43.0	0	0.0	-	-	-
pH	8	0.0	-	-	-	9	33.3	3.98	3.90	- 2.0
Ammonia-N, % TN	0	0.0	-	-	-	4	0.0	-	-	-
LAB, cfu/g	4	50.0	5.53	6.83	+ 23.5	0	0.0	-	-	-
Yeasts, cfu/g	4	0.0	-	-	-	8	50.0	6.70	2.51	- 62.5
Molds, cfu/g	0	0.0	-	-	-	8	0.0	-	-	-
Aerobic stability, h	6	16.7	42.00	54.00	+ 28.6	8	87.5	65.90	158.37	+ 140.3

[1]DM, dry matter; CP, crude protein; EE, ether extract; ADIN, acid detergent insoluble N; NDF, neutral detergent fiber; ADF, acid detergent fiber; TN, total nitrogen; LAB, lactic-acid bacteria.

Table 12. Summary of positive responses of silage inoculants on the fermentation patterns, nutritive value, and aerobic stability of high-moisture corn silages (data are given in % of DM, unless otherwise stated).

It was not observed significant differences for lactic acid production and final pH by homolactic inoculation in HMC silages. As described earlier, hoLAB are used with the goal to increase lactic acid production and quickly reduce pH of the ensiled crop [6, 8]. In addition, there is an expected inhibition on the growth of undesirable microorganisms such as enterobacteria and clostridia [6, 8]. These effects likely help us to understand why DM losses decreased by 20.4% due to homolactic inoculation. Considering the overall mean, homolactic inoculation reduced the aerobic stability by 6.9 h, compared with untreated silage. Homolactic inoculation can impair the aerobic stability of silages in some cases [32], because the lactic acid produced and the increased preservation of the forage crop can lead to an increase in the number of spoilage microorganisms, mainly yeasts.

Considering the overall mean, heterolactic inoculation of HMC silages increased the concentration of lactic and acetic acids by 106.5 and 92.7%, respectively. Due to the antifungal properties of acetic acid [30], the aerobic stability of HMC silages inoculated with ^{he}LAB increased by 102.5 h compared to untreated silage. Furthermore, heterolactic inoculation reduced the NDF content (−56%) and increased the CP content (+32%).

The frequency and difference of the positive responses found in HMC silages from homolactic and heterolactic inoculations are given in **Table 12**. Homolactic inoculation had the greatest frequency of positive responses for DM losses and LAB count. Furthermore, the greatest difference of responses was observed for DM losses and ADF content. Despite heterolactic inoculation, the greatest frequency of positive responses and the greatest magnitude of responses were observed for aerobic stability.

The fermentation of HMC silages is often restricted due to low moisture and fermentable sugar content, and the quantity of total acids produced is quite low [46]. Indeed, the data from this survey showed an increase in fermentation products in HMC silages treated with bacterial inoculants, and ^{he}LAB had the greatest impact on fermentation end products and aerobic stability.

Even without statistical analysis, the mean and median values for most variables were very similar, indicating that the data were normally distributed. Although the results of the current survey for all crops investigated are encouraging, some caution should be used when interpreting the data, because the inoculants, application rate, strains, and crops were not the same in each study and the conditions were highly variable. Moreover, the goal of this chapter was to conduct a survey that provides an exploratory picture of the silage trials carried out in Brazil, more than a proper comparison among treatments, which require analyses more specific.

2.2. Animal performance

Considerable efforts have been devoted to understand how silage inoculants affect animal performance, since such improvements are, in many cases, the principal economic justification for their use, in addition to improved nutrient recovery and enhanced aerobic stability already presented above.

Significant improvements on the performance of animals fed inoculated silages have been found in studies carried out in Europe and North America, although less frequently than studies regarding changes in fermentation caused by inoculation [47]. In a previous review concerning bacterial inoculants in Brazil (see [5]), there was not a definitive conclusion regarding the effect of inoculation on animal performance due to the low number of studies, but the authors suggested that the difference and frequency of responses should be similar to those observed in other countries (see [48]).

In our survey, we found 42 studies that included feeding inoculated silages to animals in Brazil. In these studies, feed intake, digestibility, and/or growth performance were measured. Twenty of the 42 studies were conducted in cattle, 19 in sheep, 2 in pigs, and 1 in poultry. In this survey, we summarized data into two groups of silages: (1) untreated and (2) inoculated (regardless of the type of bacterial inoculant used). Only the performance of cattle and sheep fed corn, grass, and sugarcane silages were reported in this chapter, because there were a greater number of trials in these crops than others. Nevertheless, the number of studies is much lower than those reported in the international literature.

Item[1]	Cattle												Sheep											
	Untreated						Inoculated						Untreated						Inoculated					
	n[2]	Mean	Median	SD[3]	Min	Max	n	Mean	Median	SD	Min	Max	n	Mean	Median	SD	Min	Max	n	Mean	Median	SD	Min	Max
Intake, kg/day																								
DM	4	8.37	8.08	1.53	6.63	10.70	5	7.93	7.14	1.34	6.63	10.20	4	1.06	1.07	0.20	0.73	1.37	7	1.14	1.11	0.18	0.70	1.37
DM, % BW	2	2.19	2.19	0.21	1.98	2.40	2	2.22	2.22	0.18	2.04	2.40	2	2.55	2.55	0.27	2.28	2.81	3	2.42	2.39	0.22	2.13	2.75
OM	3	7.74	6.81	1.71	6.10	10.30	4	7.22	6.61	1.29	5.87	9.80	2	1.14	1.14	0.18	0.96	1.31	4	1.09	1.04	0.08	1.02	1.25
NDF	3	2.93	2.92	0.32	2.46	3.40	4	2.84	2.84	0.26	2.40	3.30	2	0.47	0.47	0.06	0.41	0.53	4	0.48	0.47	0.03	0.44	0.51
CP	3	0.91	0.80	0.19	0.74	1.20	4	0.87	0.78	0.17	0.71	1.20	2	0.14	0.14	0.01	0.13	0.15	4	0.14	0.14	0.01	0.13	0.15
Intake of digestible nutrients, kg/day																								
DM	2	4.33	4.33	0.11	4.22	4.44	3	4.49	4.52	0.35	3.97	4.98	3	0.67	0.72	0.14	0.47	0.83	5	0.66	0.71	0.10	0.45	0.80
OM	2	4.41	4.41	0.23	4.18	4.64	3	4.44	4.56	0.34	3.92	4.83	2	0.76	0.76	0.06	0.69	0.82	4	0.70	0.70	0.05	0.60	0.79
NDF	2	1.37	1.37	0.03	1.34	1.40	3	1.36	1.35	0.13	1.17	1.55	2	0.20	0.20	0.02	0.19	0.22	4	0.21	0.21	0.02	0.19	0.24
CP	2	0.47	0.47	0.05	0.42	0.52	3	0.46	0.46	0.03	0.42	0.50	2	0.09	0.09	0.00	0.09	0.10	4	0.09	0.09	0.01	0.07	0.10
Digestibility, %																								
DM	3	64.45	63.67	1.58	62.86	66.83	4	66.49	68.18	3.33	59.83	69.77	5	62.08	60.40	4.92	55.46	71.90	7	62.23	61.20	3.79	54.80	68.30
OM	2	68.31	68.31	0.14	68.17	68.45	3	69.66	70.70	1.90	66.81	71.47	2	67.35	67.35	5.15	62.20	72.50	4	64.48	65.50	4.58	56.90	70.00
NDF	2	51.46	51.46	5.46	46.00	56.92	3	50.38	48.85	2.08	48.80	53.50	5	45.43	50.63	8.29	34.95	54.20	7	46.90	48.95	5.50	37.21	54.90
CP	2	60.93	60.93	4.40	56.53	65.33	3	60.84	60.01	1.73	59.08	63.44	3	62.37	60.50	6.55	54.41	72.20	5	62.90	66.00	4.24	56.02	66.80
Performance																								
Feed efficiency	4	6.22	6.56	1.15	4.34	7.40	4	7.12	7.14	0.65	5.90	8.28	1	-	-	-	-	-	2	5.23	5.23	0.09	5.14	5.31
ADG, kg/day	4	1.41	1.29	0.25	1.15	1.90	4	1.37	1.32	0.22	1.03	1.80	1	-	-	-	-	-	2	0.20	0.20	0.00	0.20	0.21

[1]DM, dry matter; BW, body weight; OM, organic matter; NDF, neutral detergent fiber; CP, crude protein; ADG, average daily gain.
[2]Number of means.
[3]Standard deviation.

Table 13. Range of feed intake, digestibility, and growth performance of cattle and sheep fed untreated and inoculated corn silages.

The inoculation of corn silage slightly depressed DM intake, feed efficiency, and average daily gain (ADG) of cattle (**Table 13**). Conversely, cattle fed inoculated corn silages had small increases in DM and OM digestibility, resulting in a higher intake of digestible DM (+0.16 kg/day). Regarding the performance of sheep, the inoculation of corn silage increased DM intake by 7.2%, but the digestibility and intake of digestible nutrients were unaffected, in general. Data regarding ADG were not considered, because only one study measured this parameter and, as a prerequisite of this survey, all comparisons between treatments were made considering a minimum of two studies.

The inoculation of tropical grass silages reduced the DM intake (−0.14 kg/day) in cattle (**Table 14**).

However, cattle fed inoculated grass silages exhibited better feed efficiency than cattle fed untreated silage, whereas ADG was similar between treatments (**Table 14**). Digestibility of DM, OM, NDF, and CP was slightly affected by inoculation. Furthermore, sheep fed inoculated silages exhibited higher DM intake (+11.7%), whereas bacterial inoculants had little effect on silage digestibility.

The inoculation of sugarcane silages negatively impacted DM intake in cattle (−0.56 kg/day), as well as the intake of digestible nutrients (**Table 15**). As consequence, the ADG of cattle fed inoculated silages was lower than cattle fed untreated silages (1.17 vs. 1.21 kg/day, respectively). Few measurements were made in sheep fed sugarcane silages, but positive responses from inoculation were observed on DM and NDF intake, which increased by 4.6 and 11.3%, respectively; however, inoculation reduced DM digestibility by 16.6%.

Item[1]	Cattle												Sheep											
	Untreated						Inoculated						Untreated						Inoculated					
	n[2]	Mean	Median	SD[3]	Min	Max	n	Mean	Median	SD	Min	Max	n	Mean	Median	SD	Min	Max	n	Mean	Median	SD	Min	Max
Intake, kg/day																								
DM	3	8.10	9.70	2.26	4.71	9.90	3	7.96	9.10	2.19	4.67	10.10	2	0.77	0.77	0.33	0.44	1.10	3	0.86	1.02	0.29	0.42	1.13
DM, % BW	4	2.39	2.35	0.07	2.31	2.53	4	2.36	2.38	0.05	2.25	2.42	3	1.74	1.63	0.28	1.43	2.16	6	1.90	1.89	0.23	1.36	2.29
Intake of digestible nutrients, kg/day																								
DM	3	4.84	5.31	1.16	3.10	6.13	3	4.77	5.09	1.11	3.10	6.11	2	0.50	0.50	0.29	0.21	0.78	3	0.55	0.65	0.22	0.21	0.79
Digestibility, %																								
DM	3	60.78	61.90	4.05	54.70	65.74	3	60.92	60.50	3.62	55.90	66.35	6	58.53	60.45	9.08	42.79	71.10	11	59.44	60.88	6.09	49.47	69.50
OM	2	60.10	60.10	3.40	56.70	63.50	2	60.10	60.10	2.30	57.80	62.40	1	-	-	-	-	-	2	65.25	65.25	0.20	65.04	65.45
NDF	3	50.09	44.70	8.46	42.80	62.78	3	48.80	43.50	8.40	41.50	61.39	5	55.80	54.71	8.07	44.52	69.20	9	55.99	60.67	8.77	36.10	69.48
CP	3	58.83	55.90	4.97	54.30	66.28	3	56.60	52.30	6.74	50.80	66.71	6	62.17	65.72	9.24	43.55	73.96	11	64.12	64.61	5.38	45.85	75.80
Performance																								
Feed efficiency	3	8.21	8.21	0.40	7.62	8.82	3	7.85	8.12	0.56	7.00	8.42	0	-	-	-	-	-	0	-	-	-	-	-
ADG, kg/day	3	1.15	1.10	0.10	1.06	1.30	3	1.17	1.20	0.10	1.02	1.30	0	-	-	-	-	-	0	-	-	-	-	-

[1]DM, dry matter; BW, body weight; OM, organic matter; NDF, neutral detergent fiber; CP, crude protein; ADG, average daily gain.
[2]Number of means.
[3]Standard deviation.

Table 14. Range of feed intake, digestibility, and growth performance of cattle and sheep fed untreated and inoculated grass silages.

Overall means of this survey consistently appointed for a reduction in DM intake when cattle were fed inoculated corn, grass, and sugarcane silages. However, effects of silage inoculants on feed intake and growth performance are widely varied and likely are microorganisms and strains specific along with dose dependent.

We also calculated the frequency and difference of positive responses, in addition to the impact of bacterial inoculation in experiments with cattle and sheep (**Tables 16** and **17**). There was great frequency of positive responses of inoculation concerning DM and OM digestibility in cattle fed corn silage. Similarly, inoculation had a great impact on the performance of cattle fed sugarcane silage, with feed efficiency and ADG improving by 80%. The greater ADG observed in cattle fed sugarcane silage likely arises from a better preservation of WSC during fermentation leading to the improved nutritive value of inoculated silages. In this regard, improvements in nutritive value of silages from bacterial inoculation may be strongly correlated with enhanced animal performance [47, 48]. However, the great frequency and difference of the responses might be associated with the low number of studies carried out that evaluated animal performance in Brazil.

The frequency of positive responses observed in sheep consuming inoculated silages was greater than those found in cattle. Sheep fed corn silage had a great frequency of positive responses for inoculation concerning DM, OM, NDF, and CP intake (≥50%). The ADG also improved in 50% of treatments, an overall increase of 4%. For grass silage, the greater frequency of positive responses from inoculation was observed for digestibility (DM, NDF, and CP). Conversely, only the intake of digestible NDF and NDF digestibility had positive responses by inoculation in sugarcane silages.

Item[1]	Cattle												Sheep											
	Untreated						Inoculated						Untreated						Inoculated					
	n[2]	Mean	Median	SD[3]	Min	Max	n	Mean	Median	SD	Min	Max	n	Mean	Median	SD	Min	Max	n	Mean	Median	SD	Min	Max
Intake, kg/day																								
DM	8	10.36	11.15	1.94	6.90	12.71	11	9.80	9.61	1.64	6.89	12.80	2	1.42	1.42	0.03	1.39	1.45	2	1.49	1.49	0.16	1.33	1.64
DM, % BW	4	2.16	2.33	0.44	1.29	2.70	5	2.02	2.35	0.60	1.26	2.80	1	-	-	-	-	-	1	-	-	-	-	-
OM	2	11.60	11.60	0.10	11.50	11.70	2	11.65	11.65	0.75	10.90	12.40	1	-	-	-	-	-	1	-	-	-	-	-
NDF	5	5.74	6.11	0.81	3.72	6.40	6	5.10	5.51	0.88	3.78	6.03	2	0.67	0.67	0.06	0.61	0.72	2	0.74	0.74	0.00	0.74	0.74
CP	3	1.70	1.80	0.20	1.41	1.90	3	1.65	1.80	0.27	1.25	1.90	0	-	-	-	-	-	0	-	-	-	-	-
Intake of digestible nutrients, kg/day																								
DM	3	5.62	5.45	0.94	4.37	7.02	4	4.99	4.93	0.55	4.39	5.70	1	-	-	-	-	-	1	-	-	-	-	-
OM	2	6.12	6.12	0.41	5.72	6.53	2	5.78	5.78	0.09	5.69	5.88	1	-	-	-	-	-	1	-	-	-	-	-
NDF	3	1.77	1.78	0.13	1.57	1.96	4	1.54	1.60	0.46	0.89	2.09	1	-	-	-	-	-	1	-	-	-	-	-
CP	2	1.23	1.23	0.06	1.17	1.29	2	1.13	1.13	0.02	1.11	1.15	0	-	-	-	-	-	0	-	-	-	-	-
Digestibility, %																								
DM	3	54.97	55.30	5.84	46.20	63.40	4	55.10	55.50	8.60	44.50	64.90	2	61.32	61.32	7.68	53.64	69.00	3	51.12	46.67	13.99	34.58	72.10
OM	3	57.07	55.80	5.76	49.70	65.70	4	57.98	58.85	8.18	47.40	66.80	1	-	-	-	-	-	1	-	-	-	-	-
NDF	3	35.47	29.20	11.49	24.50	52.70	4	35.95	36.90	17.00	14.80	55.20	1	-	-	-	-	-	1	-	-	-	-	-
CP	2	66.55	66.55	1.55	65.00	68.10	2	61.25	61.25	2.75	58.50	64.00	0	-	-	-	-	-	0	-	-	-	-	-
Performance																								
Feed efficiency	3	8.16	9.37	1.64	5.71	9.41	5	8.06	8.43	0.78	6.45	9.15	1	-	-	-	-	-	1	-	-	-	-	-
ADG, kg/day	3	1.21	0.94	0.44	0.82	1.87	5	1.17	1.04	0.20	0.97	1.61	1	-	-	-	-	-	1	-	-	-	-	-

[1]DM, dry matter; BW, body weight; OM, organic matter; NDF, neutral detergent fiber; CP, crude protein; ADG, average daily gain.
[2]Number of means.
[3]Standard deviation.

Table 15. Range of feed intake, digestibility, and growth performance of cattle and sheep fed untreated and inoculated sugarcane silages.

The results found in Brazilian studies suggest a greater effect of inoculation when there is a positive response, compared to those from other countries. In Europe, a review of 14 studies reported increases in DM intake (+4.8%) and milk production (+4.6%) when animals were fed silage inoculated with *L. plantarum* strain MTD1 [49]. Similarly, a review of studies carried out between 1990 and 1995 reported that in 28, 53, and 47% of these studies, there were increases in DM intake (+4.8%), ADG (+4.6%), and milk production (+4.6%), respectively [48].

Item[1]	Number of treatments		Mean		Difference, %
	Total	Positive responses, %	Untreated	Inoculated	
Corn					
Intake of digestible nutrients, kg/day					
DM	3	33.3	4.44	4.98	+12.3
OM	3	33.3	4.64	4.83	+4.0
Digestibility, %					
DM	4	75.0	64.85	68.71	+6.0
OM	3	66.7	68.17	71.09	+4.3
Sugarcane					
Intake, kg/day					
DM	11	27.3	7.69	8.74	+13.7
Intake of digestible nutrients, kg/day					
NDF	4	25.0	1.96	2.09	+6.4
Digestibility, %					
NDF	4	25.0	52.70	55.20	+4.7
Performance					
Feed efficiency	5	80.0	9.39	8.46	–9.9
ADG, kg/day	5	80.0	0.88	1.06	+21.1

[1]DM, dry matter; OM, organic matter; NDF, neutral detergent fiber; ADG, average daily gain.

Table 16. Summary of positive responses of silage inoculants on the performance of cattle fed corn and sugarcane silages in experiments carried out in Brazil.

The results of the current survey are encouraging regarding the impact of bacterial inoculants on animal performance in tropical conditions. However, although the mean and median values for most variables measuring animal performance were very similar (which may indicate normal distribution of the data), this occurred because of the lack and/or low number of studies evaluated. Therefore, some caution should be taken when interpreting this data, as well as the great frequency of positive responses found, which is likely attributed to the low number of studies evaluated.

Regarding the factors responsible for enhancing animal performance, certainly improvements in DM digestion are closely linked to greater growth performance. In a review of the literature from 1985 to 1992, animal performance improved in 9 of 16 trials when inoculation improved DM digestion, but only 2 of 15 trials when digestion was not significantly affected [50].

In our survey, we did not observe a relationship between DM digestibility and growth performance, because the number of studies evaluated was quite low. However, there are other hypotheses related to the improvement of animal performance. The first suggests that improvements in silage quality could lead to increased animal performance. The second suggests that silage inoculants may provide a probiotic effect by inhibiting detrimental microorganisms in the silage and rumen, or by producing beneficial substances that may enhance the functioning of specific microbial populations in the rumen, leading to an increase in animal performance [47].

Item[1]	Number of treatments		Mean		Difference, %
	Total	Positive responses, %	Untreated	Inoculated	
Corn					
Intake, kg/day					
DM	7	57.1	1.07	1.21	+ 13.0
OM	4	50.0	0.96	1.02	+ 6.4
NDF	4	50.0	0.41	0.44	+ 8.0
CP	4	50.0	0.13	0.14	+ 7.8
Digestibility, %					
CP	5	20.0	60.50	66.00	+ 9.1
Performance					
ADG, kg/day	2	50.0	0.20	0.21	+ 4.0
Grass					
Intake, % BW					
DM	6	50.0	1.43	1.97	+ 38.0
Digestibility, %					
DM	11	36.4	49.95	54.14	+ 8.4
NDF	9	33.3	54.71	60.64	+ 10.8
CP	11	9.1	53.07	68.76	+ 29.6
Sugarcane					
Intake of digestible nutrients, kg/day					
NDF	1	100.0	0.35	0.48	+ 36.9
Digestibility, %					
NDF	1	100.0	57.50	64.90	+ 12.9

[1]DM, dry matter; OM, organic matter; NDF, neutral detergent fiber; CP, crude protein; ADG, average daily gain; BW, body weight.

Table 17. Summary of positive responses of silage inoculants on the performance of sheep fed corn, grass, and sugarcane silages in experiments carried out in Brazil.

A probiotic can be defined as a culture of live microbes, that when fed to the animals, beneficially affects the host by improving the properties of the native gut microflora [48]. Indeed, a recent study displayed greater microbial protein synthesis in lambs fed silage inoculated with *L. buchneri,* applied either alone or associated with *L. plantarum* in corn silage [51], which is likely related to changes in the microbial community in the rumen.

3. Implications

The data summarized from Brazilian studies displays a recent increase in interest from researchers addressing bacterial inoculants as an alternative to improve silage quality. But although the number of studies remains quite low compared with the international literature, data of this survey revealed some trends for improved fermentation and nutritive value regarding the group of bacterial inoculant used at ensiling and crop.

Considering an overall mean, homolactic inoculation unaffected DM losses in corn, grass, HMC, and sorghum silages, but reduced DM loss in alfalfa silages. However, an unexpected increase in aerobic stability of grass silage was reported due to homolactic inoculation. The greater frequency of positive response was also observed for grass silages when treated with ^{ho}LAB. Conversely, heterolactic inoculation revealed to be more interesting than homolactic inoculants to reduce fermentation losses in sugarcane silage, and positive responses were found most often. In addition, enhanced aerobic stability was reported for corn and HMC silages when they were treated with ^{he}LAB. Overall, the results of the current survey regarding fermentation patterns of inoculated silages are encouraging, mainly for grass and sugarcane silages. Otherwise, the impact of bacterial inoculant on silage quality (i.e., fermentation patterns, chemical composition, and nutritive value) appeared to diminish as the quality of ensiled crop increased.

Despite of animal performance and considering the overall means, inoculation consistently depressed DM intake in cattle fed corn, grass, and sugarcane silages, but DM intake increased in sheep due to inoculation. There were not a consistent effect of bacterial inoculants on silage digestibility, which largely varied depending the animal and crop evaluated. Conversely, cattle fed inoculated sugarcane silage had a greater frequency of positive response on ADG. The performance of animals consuming inoculated silages has been investigated in Brazil only a few times, but the data suggest a greater impact of bacterial inoculants on DM intake and weight gain in cattle and sheep than that indicated under temperate conditions. However, the number of studies evaluating animal performance still remains quite low, especially for dairy cows fed inoculated silage, and this survey did not provide a definitive conclusion about the effect of bacterial inoculants on animal performance (cattle and sheep).

Finally, we need caution to interpret the data of the current survey because the potential of bacterial inoculants measured by studies containing positive responses were highly variable and deeply associated with number of studies. Hence, a greater frequency of positive responses was often observed when there were a low number of studies evaluated. Additionally, positive responses were clearly impacted by the group of microorganisms (homo and heterofermentative LAB) and it determined the success of bacterial inoculant applications in silage. In this way, the compatibility between the plants and microorganisms used at ensiling should be taken into account in further studies, as well as its applicability on farm. In addition, further studies may consider assessing animal performance and sanitary aspects related to the use of bacterial inoculants since there is a lack of data about it.

Acknowledgements

'The authors wish to express their appreciation to Lallemand Animal Nutrition for collaborating with this survey and providing financial support' by 'The authors wish to thank the São Paulo Research Foundation (FAPESP grant #2016/04484-0) and Lallemand Animal Nutrition for collaborating with this survey and providing financial support'.

http://dx.doi.org/10.5772/64472

Author details

Carlos H.S. Rabelo[1*], Lucas J. Mari[2] and Ricardo A. Reis[1,3]

*Address all correspondence to: carlos.zoo@hotmail.com

1 Department of Animal Science, São Paulo State University, Jaboticabal, São Paulo, Brazil

2 Lallemand Animal Nutrition, Aparecida de Goiânia, Goiás, Brazil

3 Department of Animal Science, INCT/CA, Federal University of Viçosa, Viçosa, Minas Gerais, Brazil

References

[1] Bolsen KK, Ashbell G, Weinberg, ZG. Silage fermentation and silage additives – review. Asian-Australasian Journal of Animal Sciences. 1996;9:483–494.

[2] Costa JHC, Hotzel MJ, Longo C, Balcao LF. A survey of management practices that influence production and welfare of dairy cattle on family farms in southern Brazil. Journal of Dairy Science. 2013;96:307–317.

[3] Oliveira CA, Millen DD. Survey of the nutritional recommendations and management practices adopted by feedlot cattle nutritionists in Brazil. Animal Feed Science and Technology. 2014;197:64-75.

[4] Bernardes TF, Rêgo AC. Study on the practices of silage production and utilization on Brazilian dairy farms. Journal of Dairy Science. 2014;97:1852-1861.

[5] Zopollatto M, Daniel JLP, Nussio LG. Microbial silage additives in Brazil: review of aspects of ensilage and animal performance. Brazilian Journal of Animal Science. 2009;38:170-189.

[6] McDonald P, Henderson AR, Heron SJE. The Biochemistry of Silage. 2nd ed. Chalcombe Publications, Abersytwyth; 1991. 340 p.

[7] Rooke JA, Hatfield RD. Biochemistry of ensiling. In: Buxton DR, Muck RE, Harrison JH, editors. Silage Science and Technology. American Society of Agronomy, Madison, WI; 2003. p. 95–140.

[8] Pahlow G, Muck RE, Driehuis F, Oude-Elferink SJWH, Spoelstra SF. Microbiology of ensiling. In: Buxton DR, Muck RE, Harrison JH, editors. Silage Science and Technology. American Society of Agronomy, Madison, WI; 2003. p. 31–93.

[9] Kung LJr, Stokes MR, Lin CJ. Silage additives. In: Buxton DR, Muck RE, Harrison JH, editors. Silage Science and Technology. American Society of Agronomy, Madison, WI; 2003. p. 305-360.

[10] Wilkinson JM, Toivonen MI. World Silage: A Survey of Forage Conservation Around the World. Chalcombe Publications; 2003; Southampton, UK.

[11] Muck RE. Silage microbiology and its control through additives. Brazilian Journal of Animal Science. 2010;39:183-191.

[12] Holzer M, Mayrhuber E, Danner H, Braun R. The role of *Lactobacillus buchneri* in forage preservation. Trends in Biotechnology. 2003;21:282-287.

[13] Filya I. The effect of *Lactobacillus buchneri* and *Lactobacillus plantarum* on the fermentation, aerobic stability, and ruminal degradability of low dry matter corn and sorghum silages. Journal of Dairy Science. 2003;86:3575-3581.

[14] Reich LJ, Kung L Jr. Effects of combining *Lactobacillus buchneri* 40788 with various lactic acid bacteria on the fermentation and aerobic stability of corn silage. Animal Feed Science and Technology. 2010;159:105-109.

[15] Filya I. The effect of *Lactobacillus buchneri*, with or without homofermentative lactic acid bacteria, on the fermentation, aerobic stability and ruminal degradability of wheat, sorghum and maize silages. Journal of Applied Microbiology. 2003;95:1080-1086.

[16] Bernardes TF, Adesogan AT. Aerobic deterioration of silages in warm climates. In: Proceedings of the 6th Symposium on Strategic Management of Pastures; Viçosa, 2012. p. 249-268.

[17] Santos AO, Ávila CLS, Schwan RF. Selection of tropical lactic acid bacteria for enhancing the quality of maize silage. Journal of Dairy Science. 2013;96:7777-7789.

[18] Fugita CA, Prado IN, Jobim CC, Zawadzki F, Valero MV, Pires MCO, Prado RM, Françozo MC. Corn silage with and without enzyme-bacteria inoculants on performance, carcass characteristics and meat quality in feedlot finished crossbred bulls. Brazilian Journal of Animal Science. 2012;41:154-163.

[19] Hoffman PC, Esser NM, Shaver RD, Coblentz WK, Scott MP, Bodnar AL, Schmidt RJ, Charley RC. Influence of ensiling time and inoculation on alteration of the starch-protein matrix in high-moisture corn. Journal of Dairy Science. 2011;94:2465-2474.

[20] Owens FN, Zinn RA, Kim YK. Limits to starch digestion in the ruminant small intestine. Journal of Animal Science. 1986;63:1634-1648.

[21] Freitas AWP, Pereira JC, Rocha FC, Costa MG, Leonel FP, Ribeiro MD. Evaluation of the nutritional quality of sugarcane silage treated with microbial additives and soybean crop residue. Brazilian Journal of Animal Science. 2006;35:38-47.

[22] Muck, RE. 2008. Advances in Inoculants for Silage. Presented at the IV Symposium on Strategic Management of Pasture, Vicosa, Minas Gerais, Brazil. November 13-15. Pereria OG, Obeid JA, da Fonseca DM, do Nascimento Jr., D, editors. Pg 221-232.

[23] Kung LJr, Shaver R. Interpretation and use of silage fermentation analysis reports. Focus on Forage. 2001;3:1-5.

[24] Ashbell G, Weinberg ZG, Hen Y, Filya I. The effects of temperature on the aerobic stability of wheat and corn silages. Journal of Industrial Microbiology & Biotechnology. 2002;28:261-263.

[25] Oude Elferink SJWH, Krooneman J, Gottschal JA, Spoelstra SF, Faber F. Anaerobic conversion of lactic acid to acetic acid and 1,2-propanediol by *Lactobacillus buchneri*. Applied and Environmental Microbiology. 2001;67:125-132.

[26] Kang TW, Adesogan AT, Kim SC, Lee SS. Effects of an esterase-producing inoculant on fermentation, aerobic stability, and neutral detergent fiber digestibility of corn silage. Journal of Dairy Science. 2009;92:732-738.

[27] Nsereko VL, Smiley BK, Rutherford WM, Spielbauer A, Forrester KJ, Hettinger GH, Harman EK, Harman BR. Influence of inoculating forage with lactic acid bacterial strains that produce ferulate esterase on ensilage and ruminal degradation of fiber. Animal Feed Science and Technology. 2008;145:122-135.

[28] Kung L Jr, Ranjit NK. The effect of *Lactobacillus buchneri* and other additives on the fermentation and aerobic stability of barley silage. Journal of Dairy Science. 2001;84:1149-1155.

[29] Kleinschmit DH, Kung LJr. A meta-analysis of the effects of *Lactobacillus buchneri* on the fermentation and aerobic stability of corn and grass and small-grain silages. Journal of Dairy Science. 2006;89:4005-4013.

[30] Moon N. Inhibition of the growth of acid tolerant yeasts by acetate, lactate and propionate and their synergistic mixtures. Journal of Applied Bacteriology. 1983;55:453-460.

[31] Meeske R, Basson HM, Cruywagen CW. The effect of lactic acid and bacterial inoculant with enzymes on the fermentation dynamics, intake and digestibility of *Digitaria eriantha* silage. Animal Feed Science and Technology. 1999;81:237-248.

[32] Adesogan AT. Avoiding the two greatest silage problems. In: Proceedings 50th Florida Dairy Production Conference; 9 April 2014; Gainesville. p. 9-17.

[33] Jonsson A. The role of yeasts and clostridia in silage deterioration—identification and ecology [dissertation]. Uppsala: Swedish University of Agricultural Sciences; 1989.

[34] Lin C, Bolsen KK, Brent BE, Hart RA, Dickerson JT, Feyerherm AM, Aimutis WR. Epiphytic microflora on alfalfa and whole-plant com. Journal of Dairy Science. 1992;75:2484-2493.

[35] Ávila CLS, Pinto JC, Figueiredo HCP, Schwan RF. Effects of an indigenous and a commercial *Lactobacillus buchneri* strain on quality of sugar cane silage. Grass and Forage Science. 2009;64:384-394.

[36] Axelsson L. Lactic acid bacteria: classification and physiology. In: Salminen S, Von Wright A, editors. Lactic Acid Bacteria: Microbiology and Functional Aspects. Marcel Dekker, New York, NY; 1998. p. 1-72.

[37] Driehuis F, Oude Elferink SJWH, Spoelstra SF. Anaerobic lactate degradation in maize silage inoculated with *Lactobacillus buchneri* inhibits yeast growth and improves aerobic stability. Journal of Applied Microbiology. 1999;87:583-594.

[38] Krooneman J, Faber F, Alderkamp AC, Oude-Elferink SJWH, Driehuis F, Cleenwerck I, Swings J, Gottschal JC, Vancanneyt M. *Lactobacillus diolivorans* sp. *Nov.*, a 1,2-propanediol-degrading bacterium isolated from aerobically stable maize silage. International Journal of Systematic and Evolutionary Microbiology. 2002;52:639-646.

[39] Danner H, Holzer M, Mayrhuber E, Braun R. Acetic acid increases stability of silage under aerobic conditions. Applied and Environmental Microbiology, 2003;69:562-567.

[40] Fitzsimons A, Duffner F, Curtin D, Brophy G, O'Kiely P, O'Connell M. Assessment of *Pediococcus acidilactici* as a potential silage inoculant. Applied and Environmental Microbiology. 1992;58:3047-3052.

[41] Daniel JLP, Checolli M, Zwielehner J, Junges D, Fernandes J, Nussio LG. The effects of *Lactobacillus kefiri* and *L. brevis* on the fermentation and aerobic stability of sugarcane silage. Animal Feed Science and Technology. 2015;205:69-74.

[42] Cai Y, Kumai S, Zhang J, Benno, Y. Comparative studies of Lactobacilli and Enterococci associated with forage crops as silage inoculants. Animal Science Journal. 1999;70:188-194.

[43] Kung LJr, Grieve DB, Thomas JW, Huber JT. Added ammonia or microbial inocula for fermentation and nitrogenous compounds of alfalfa ensiled at various percents of dry matter. Journal of Dairy Science. 1984;67:299-306.

[44] Muck RE. Initial bacteria numbers on lucerne prior to ensiling. Grass and Forage Science. 1989;44:19-25.

[45] Bolsen KK, Moore KJ, Coblentz WK, Siefers MK, White JS. Sorghum silage. In: Buxton DR, Muck RE, Harrison JH, editors. Silage Science and Technology. American Society of Agronomy, Madison, WI; 2003. p. 609-632.

[46] Kung LJr, Schimidt RJ, Ebling TE, Hu W. The effect of *Lactobacillus buchneri* 40788 on the fermentation and aerobic stability of ground and whole high-moisture corn. Journal of Dairy Science. 2007;90:2309-2314.

[47] Weinberg ZG, Muck RE. New trends and opportunities in the development and use of inoculants for silage. FEMS Microbiology Reviews. 1996;19:53-68.

[48] Kung LJr., Muck RE. Animal response to silage additives. In: Silage: Field to Feedbunk. Northeast Regional Agricultural Engineering Service, Ithaca, NY; 1997. p. 200-210.

[49] Moran JP, Owen TR. The effects of Ecosyl treated silage on milk production by lactating cows. In: Proceedings of the National Conference on Forage Quality, Evaluation and Utilization. Lincoln. University of Nebraska; 1994. p. 126.

[50] Muck RE. The role of silage additives in making high quality silage. In: Proceedings of the National Silage Production Conference; New York, 23–25 February 1993; NRAES, Ithaca, NY. p. 106-116

[51] Basso FC, Adesogan AT, Lara EC, Rabelo CHS, Berchielli TT, Teixeira IAMA, Siqueira GR, Reis RA. Effects of feeding corn silage inoculated with microbial additives on the ruminal fermentation, microbial protein yield, and growth performance of lambs. Journal of Animal Science. 2014;92:5640-5650.

Chapter 2

Environmental Factors Affecting Corn Quality for Silage Production

Gonzalo Ferreira and Alston N. Brown

Additional information is available at the end of the chapter

http://dx.doi.org/10.5772/64381

Abstract

Corn silage is a major ingredient of diets for dairy cattle. Environmental factors can affect the yield and composition of corn silage. Drought and heat are two common environmental factors that affect silage yield and quality. Corn silages with low concentrations of dry matter, high concentrations of protein, high concentrations of fiber, and low concentrations of starch indicate that the crop was harvested too early, that abiotic stresses affected the structure of the plant, or a combination of both. Drought stress during vegetative stages does not affect yield and nutritional composition as much as during reproductive stages. High environmental temperatures (>35 °C) can also induce kernel abortion. The effects of abiotic stresses on cell wall composition are less clear. Drought stress would likely increase fiber digestibility, whereas heat stress would decrease fiber digestibility. These statements are somehow contradictory in the sense that drought stress and heat stress likely occur simultaneously. Management practices, such as hybrid selection and planting date, should be considered to avoid silking and early kernel development during season of very high environmental temperatures.

Keywords: corn silage, drought stress, heat stress, abiotic stress, nutritional quality

1. Introduction

Whole-plant corn silage is a major ingredient of diets fed to dairy cattle; therefore, producing high-yielding and good-quality corn silage is critical for minimizing production costs in dairy farming systems. The US dairy industry is composed of 9.2 million cows and approximately 4.5 million replacement heifers [1], which consume approximately 60 million (metric) tons of corn silage per year (**Table 1**). The high inclusion of corn silage in diets for dairy cows is attributed to multiple factors. First, corn silage is an attractive feed source because of high yield potential.

For example, dry matter (DM) yields per acre are substantially greater for corn silage than for alfalfa hay (12,600 and 7200 kg/ha, respectively) [2]. Second, corn silage is also characterized by having high concentrations of energy. Under normal climatic conditions, the corn plant contains a great proportion of starch-containing grains. This starch is highly digestible and therefore is an important source of energy for cattle. Finally, corn silage also provides fiber in ruminant diets. Dairy cows require a minimum amount of dietary fiber to ensure ruminal and whole-animal health [3].

Different crop management practices, such as planting density, nitrogen fertilization rates, harvesting time, or harvesting height, can affect corn silage yield, corn silage quality, or both [4]. One way or another, most of these factors, if not all, can be controlled based on managerial decisions. In addition to controllable factors, there are several uncontrollable environmental factors that can substantially affect the dry matter yield and the nutritional composition of corn used for whole-plant corn silage.

	Milk cows	Replacement	Total
Cow inventory, million heads	9.2	4.5	–
Corn silage consumption, kg/head/day	15.0	5.5	–
Corn silage consumption, million ton[1]/year	50.4	9.0	59.4
Corn silage price, $/ton[1]	45	45	45
Expenditure in corn silage, billion $/year	2.26	0.41	2.67

[1]Metric ton = 1,000 kg

Table 1. Consumption and expenditure for corn silage by the US dairy industry.

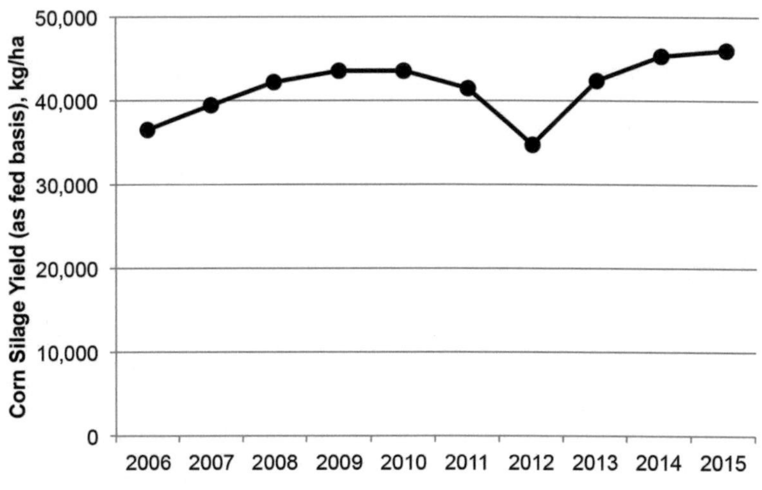

Figure 1. National US corn silage yields (kg/ha, as-fed basis). Spring and summer drought of 2012 will be remembered as one of the "worst agricultural calamities in the United States" [21].

http://dx.doi.org/10.5772/64381

Drought and heat stresses, also known as abiotic stresses, are two common and interrelated environmental factors that frequently affect corn silage yield and quality [4]. The impact of these factors can be substantial. For example, the drought of 2012 reduced US national silage yields by 16.3 % when compared to 2011 (**Figure 1**). This reduction in yield caused the United States an economic loss between \$700 and \$800 million for 2012. This loss does not take into account the overall impact to the dairy industry, such as increases in feed prices for hay and corn grain.

Even though drought stress and heat stress are uncontrollable factors that affect corn silage yield and quality, certain management practices can be utilized to attenuate their potential negative impact. The objective of this chapter is to describe such practices so that crop managers can minimize the negative effects of abiotic stresses in yield and quality of corn silage.

2. The corn plant

The corn plant is characterized by having a single erect stem that is divided into basic units known as phytomers. Each phytomer consists of a leaf blade, a leaf sheath, a node, an internode, and the axillary bud. Different from most other grasses, the corn plant has two separate inflorescences per plant, the tassel and the ear, which are the male and the female inflorescences, respectively. The husks are leaves that cover the ear, where corn kernels develop after pollination. Corn kernels are arranged and inserted in lines on an inner cylinder called the cob, which is originated from the axillary bud from the phytomers.

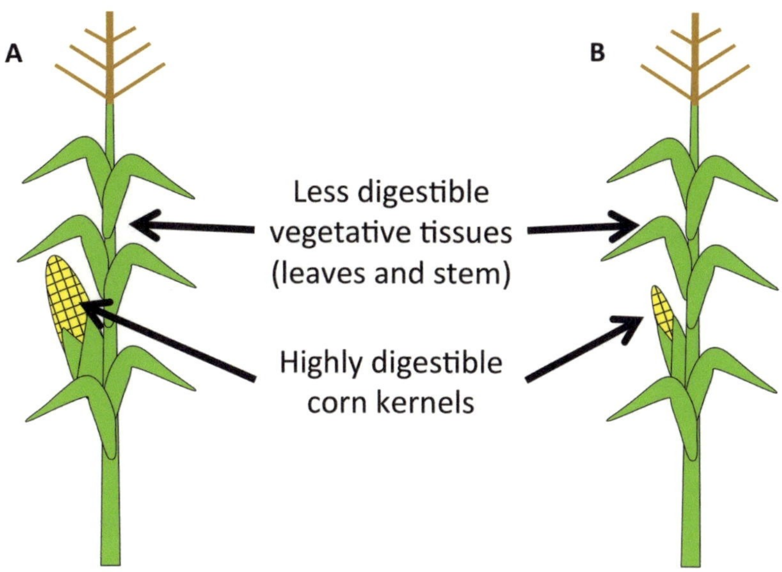

Figure 2. The proportion of grain in the corn plant has a major impact on corn silage yield and nutritional quality. The bigger ear in plant A will result in greater yields of dry matter and greater energy concentration than in plant B.

The structure of the corn plant has a major impact on the chemical composition of corn silage. Carbohydrates synthesized in leaves are mobilized to the grain and stored as starch. Corn kernels comprise 30–52 % of the total plant biomass [5], whereas starch constitutes 70–75 % of the kernel dry weight [6]. Because of the different composition of the grain and the vegetative portion of the plant (high and low concentrations of nonstructural carbohydrates, respectively), the proportion of grain in the corn plant has a substantial impact on the nutritional quality of corn silage (**Figure 2**).

In corn, inflorescence development occurs during the vegetative growth stages of the crop, typically when corn plants have six fully exposed leaves (stage known as V6) [7]. At this stage, the axillary meristem of leaves differentiates into ears [8]. These ears typically produce rows of paired spikelets (**Figure 3**) that produce one ovule-containing floret each. After pollination, when the ovule is successfully fertilized, each floret results in a single corn kernel.

The number of kernels per plant is known as the sink capacity of the plant, which is determined by three components: (1) the number of spikelet rows within the ear, (2) the number of spikelets per row, and (3) the proportion of single and double spikelets within a row (**Figure 3**). Because the sink capacity determines the potential number of kernels in the plant and because the proportion of kernels is a major determinant of the nutritional quality of the whole plant, it is likely that ear differentiation has a major impact on corn silage quality.

Unlike most other grasses, the male inflorescence is separated spatially from the female inflorescence in corn. Every ovule within the ear has to be pollinated to become a developed kernel. For this process, functional stigmas, known as silks, connect the ovules to the exterior of the ear to ensure pollination. The appearance and exposure of the silk to the environment is known as the silking stage and is considered the beginning of the reproductive stage of the corn crop. The first step in the pollination process occurs when pollen grains released from the tassel during anthesis attach to ear silks. The synchrony between anthesis and the emergence of silks (commonly known as anthesis-silking interval, ASI) is critical for adequate kernel pollination and development [9].

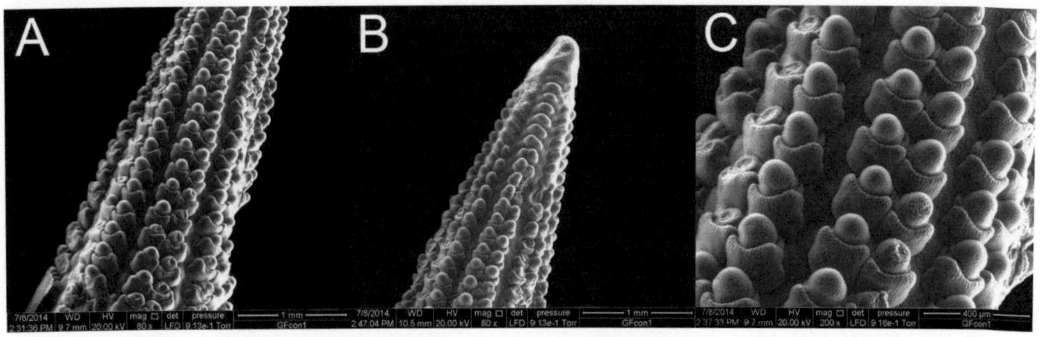

Figure 3. Scanning electron microscopy (SEM) images of corn ears during kernel differentiation. Ear differentiation in the corn plant determines the number of kernels in the whole plant. Normal plants develop row of paired spikelets (A and C), which result in corn kernels. When ear differentiation is affected, irregular rows of single spikelets (B) could be observed. Images were obtained at the Nanoscale Characterization and Fabrication Laboratory (Virginia Tech).

An understanding of corn plant composition is crucial to comprehend the effects of abiotic stresses on the composition of corn silage. In the end, kernel differentiation and kernel development and growth will determine the final number of kernels per plant and, therefore, the starch and fiber concentrations in whole-plant corn silage.

3. Nutritional quality of stressed corn silage

As an ingredient in rations for dairy cows, the value of corn silage relies mainly on its energy concentration and not so much on its crude protein concentration. For example, corn silage typically contains low concentrations of crude protein compared to alfalfa haylage (less than 10 % and more than 15 %, respectively). The low crude protein concentration of the whole corn plant is related to the structure of the corn plant. Corn grain is characterized as having low concentrations of protein [3, 6] due to the high proportion (more than 82 %) of a starchy and nonprotein endosperm. Corn kernels also contain less moisture than vegetative tissues, such as stems and leaves. Therefore, corn silages with high proportions of grain (i.e., a high harvest index) would likely have high concentrations of dry matter (>30 % dry matter), low concentrations of crude protein (<10 % crude protein), low concentrations of fiber (<45 % neutral detergent fiber), and high concentrations of starch (>30 % starch). In contrast, corn silages with relatively low concentrations of dry matter, high concentrations of crude protein, high concentrations of fiber, and low concentrations of starch reflect an indication that either the crop was harvested too early, abiotic stresses affected the structure of the corn plants, or a combination of both.

In a retrospective study performed at Virginia Tech [4], corn hybrids harvested for silage in 2 years, which included 2012, at two sites were analyzed to understand how dry matter yields and nutritional composition were affected by abiotic stresses (**Table 2**). Dry matter yields varied significantly across site-years, but not between hybrids. Even though in 2012 rainfalls were scarce and similar at both sites (262 and 227 mm for the Shenandoah Valley and Southern Piedmont, respectively), dry matter yields and nutritional composition of corn plants differed substantially among locations. Dry matter concentration was substantially low (25.3 % dry matter) in the Southern Piedmont only, likely due to a reduced proportion of the grain component in the whole plant. The low dry matter concentration was followed by a relatively high concentration of crude protein (10.9 % crude protein) and a relatively high concentration of fiber (56.6 % neutral detergent fiber). In contrast to this, dry matter (32.6–37.0 % dry matter) and crude protein (7.1–8.7 % crude protein) concentrations were within typical values for other site-years. Even though the concentrations of fiber were more variable (43.0–52.8 % neutral detergent fiber) in other site-years, these values were lower than those observed in 2012 in the Southern Piedmont. In summary, during the spring and summer drought of 2012, an evident stress was noticed by visual appraisal of corn plots in the Southern Piedmont. This stress manifested with low concentrations of dry matter and high concentrations of crude protein and fiber.

	Southern Piedmont		Shenandoah Valley	
	2011	2012	2011	2012
Planting date	April 18	April 10	May 6	May 21
Harvesting date	August 31	July 17	August 24	September 12
Rainfalls, mm	501	228	280	262
Rainfall Shannon diversity index	0.65	0.66	0.60	0.67
Dry matter yield, kg/ha	12,482	4,556	15,092	12,678
Dry matter concentration, %	37.0	25.3	32.6	35.4
Crude protein concentration, %	8.7	10.9	7.7	7.1
Neutral detergent fiber concentration, %	51.5	56.5	52.8	43.0
Data from Ferreira et al. [4]				

Table 2. Dry matter yield and nutritional composition of corn hybrids tested at two locations in Virginia (United States) during 2011 and 2012.

4. Drought stress and kernel development

Water status of the plant is determined by several factors, including the amount and distribution of rainfalls, evapotranspiration, and the water-holding capacity of the soil. The interaction between these factors can substantially affect yields and nutritional composition of corn for silage. Adequate soil moisture is critical to ensure germination and emergence of corn seedlings soon after planting. After seedling emergence, the relatively low evapotranspiration allows plants to grow with minimum stress as long as water content in the soil is adequate. For example, limiting irrigation in corn plots during vegetative stages (i.e., six-leaf stage) reduced neither the grain yield per hectare nor the number of kernels per ear when compared to corn plots receiving complementary irrigation during the vegetative stage [10]. In contrast, limiting irrigation around silking reduced the grain yield per hectare and the weight of the kernels, although the number of kernels per ear was not affected when compared to corn plots receiving complementary irrigation during vegetative stages [10]. These data suggest that when drought stress occurs at vegetative stages, dry matter yields can be compromised but kernel development and the potential nutritional composition of the silage are not necessarily affected.

Unlike in vegetative stages, drought stress during reproductive stages can substantially affect kernel development [9–11]. NeSmith and Ritchie [11] and Çakir [10] reported substantial reductions in the number of kernels per ear when corn plants were subjected to water deficits around silking stage. Although it is clear that drought stress around silking impacts kernel development, multiple mechanisms affect this process.

The seed set is determined during vegetative stages, so the number of ovaries per ear (i.e., the potential number of kernels per ear) is not greatly affected by drought stress around silking

[9]. On the other hand, ovary atrophy or abortion occurs when water stress occurs around silking, reducing kernel development and growth within the ear. Drought stress around silking retards growth and emergence of silks, especially those from apical ovaries (**Figure 4**). The delayed emergence of silks relative to anthesis increases the asynchrony between pollen shed and silking, which can potentially decrease pollination and ovule fertilization. Depending on genotypes and stress levels, drought stress can increase the anthesis-silking interval from 1.9 to 4.8 days [9]. The synchrony between anthesis and silking has become quite relevant in breeding programs, as reducing the time elapsed between anthesis and silking is the main strategy for increasing the tolerance of corn to drought stress [9, 12].

Figure 4. Drought-stressed corn crop showing poor kernel development in the apical region of the ear.

5. Heat stress and kernel development

Drought stress and heat stress tend to occur simultaneously. In general terms, high environmental temperatures will increase evapotranspiration, exacerbating the effects of drought stress, especially when it is accompanied by low relative humidity. Despite this, these two abiotic stresses may affect kernel development by different mechanisms, affecting the composition of corn silage in different ways [4].

Schoper et al. [13] evaluated the effect of drought stress and heat stress on seed set or kernel development while considering the impact of heat stress over the pollen source (i.e., the tassel). As in other studies, the number of kernels per ear decreased approximately 17–19 % when the silk source was subjected to water stress, and the magnitude of this decrease was similar when the pollen source was also subjected to water stress. This last observation indicated that the production of viable pollen was not affected by drought stress. However, when pollen source was subjected to heat stress, the number kernels per ear decreased by approximately 72 % when the silk source was well watered and by approximately 85 % when the silk source was subjected to drought stress. These observations indicated that heat stress had an adverse effect on the development of viable pollen [13], resulting in limited pollination and ovule fecundation.

In addition to limiting pollination, heat stress can limit kernel development after ovule fecundation [14, 15]. Kernel development is divided by a lag phase with little kernel growth and a linear growing phase with major accumulation of dry matter. The lag phase, which starts immediately after pollination and lasts 10 to 12 days after pollination, is critical for kernel development [15]. The endosperm is the structure of the corn kernel that contains starch granules. Cell division of the endosperm cells during the lag phase determines the capacity of the endosperm to accumulate starch within the grain [15].

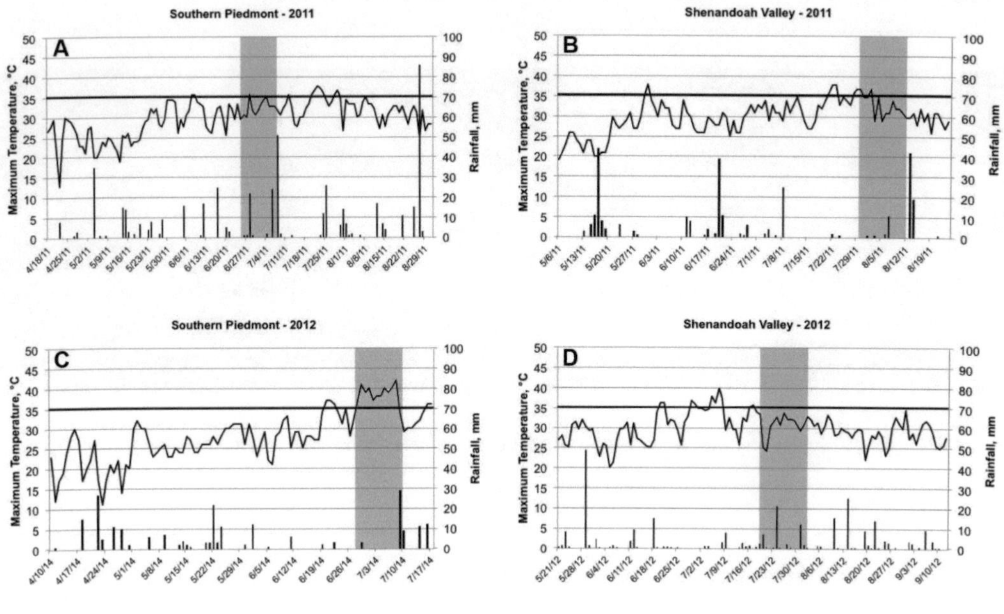

Figure 5. Daily maximum temperatures (line) and rainfalls (columns) during the crop cycle at two regions during 2011 and 2012 in the state of Virginia. The shaded region represents the critical stage for kernel development. The thick horizontal line represents the threshold temperature for heat stress (>35 °C). Prolonged heat stress after silking occurred only in the Southern Piedmont region during 2012 (C), but not in other site-years (A, B, and D). Data from Ferreira et al. [4].

High temperatures immediately after silking limit starch accumulation within the kernels and increase the rate of kernel abortion as well. Cheikh and Jones [15] cultured corn kernels in vitro at different temperatures and observed that heat-stressed kernels (i.e., kernels cultured at 35 °C) accumulated 18–75 % less DM than non-stressed kernels (i.e., kernels cultured at 25 °C). Reduced dry matter accumulation can be related to reductions in starch synthesis within the endosperm when kernels are subjected to temperatures greater than 35 °C [14]. In addition to reduced kernel growth, Cheikh and Jones [15] reported 23–97 % kernel abortion when subjected to heat stress.

In their retrospective study, Ferreira et al. [4] observed that in 2011, maximum temperatures were below 35 °C throughout the whole critical period of kernel development for the Southern Piedmont region, whereas in the Shenandoah Valley region, maximum temperatures were above 35 °C for only a few days during the critical period of kernel development (**Figure 5B**).

Based on these observations, heat stress would not have affected kernel development. In 2012, however, the Southern Piedmont region had maximum daily temperatures above 35 °C for an extended period (11 days) right after silking (**Figure 5C**), whereas maximum daily temperatures were 7.1 ± 2.3 °C lower in the Shenandoah Valley region around silking (**Figure 5D**). It is therefore likely that heat stress had a major effect on kernel development in the Southern Piedmont region but not in the Shenandoah Valley region during 2012. Therefore, in the Southern Piedmont region, heat stress exacerbated the effects of drought, substantially reducing dry matter yields and kernel development. Similar observations were reported for the southern region of the United States for 2012 [4].

In conclusion, in regions with extended periods of temperatures greater than 35 °C, choosing early maturity corn hybrids or delaying planting date should be considered to avoid drought and heat stress during silking and kernel development.

6. Abiotic stresses and cell wall composition

The effects of abiotic stresses on cell wall composition are less clear than their effects on kernel development. In general terms, and from a nutritional perspective, drought stress would likely increase fiber digestibility (**Table 3**, data Argentina), whereas heat stress would decrease fiber digestibility [16]. These statements are somehow conflicting in the sense that drought stress and heat stress likely occur simultaneously.

	2008	2009
Dry matter concentration, %	32.2	28.5
Crude protein concentration, %	8.1	7.3
Neutral detergent fiber concentration, %	45.0	49.2
Starch concentration, %	18.8	7.0
Fiber digestibility[1], %	44.4	52.6

[1]30 h neutral detergent fiber digestibility

Table 3. Nutritional composition and digestibility of corn silages in Buenos Aires (Argentina) during normal (2008) and drought (2009) years.

Drought stress during early vegetative stages can result in shorter internode lengths as a consequence of limited cell growth or elongation (**Figure 6**). As internodes contain highly lignified tissues (e.g., lignified vascular bundles), the concentration of lignin within the cell wall could be reduced when considering the whole corn plant. In addition to changes in whole plant structure (i.e., internode elongation), lignification might decrease at the tissue level when corn plants are subjected to drought stress [17, 18]. Vincent et al. [17] reported that lignin accumulation in the apical zone of corn leaves was reduced in response to drought stress. Alvarez et al. [18] reported higher concentrations of lignin precursors (i.e., p-coumaric and

Figure 6. Drought-stressed corn crop passed tassel emergence, showing reduced elongation of internodes.

caffeic acids) in xylem sap of drought-stressed corn compared to well-watered corn, suggesting reductions in lignin concentration under drought stress.

7. Abiotic stresses and silage fermentation

Because controlled experiments evaluating the effects of abiotic stresses on corn silage are scarce, most of the knowledge on silage fermentation may be obtained from field experience. One reason for the lack of controlled studies may be that accomplishing and reproducing stress treatments are difficult [19].

One major concern when ensiling stressed corn can be the low DM concentration of the forage. As described before, if poor kernel development occurs, then low DM concentrations will likely occur [4], and therefore it might be very difficult to obtain a high enough DM concentration (>30 % DM) for an adequate ensiling process. In these scenarios, the likelihood of seepage losses or clostridial fermentations may increase [20]. On the other hand, drought stress conditions might also increase solute concentrations, which could decrease water activity and growth of lactic acid bacteria [20]. In regard to silage density, packing may be more challenging with heat-stressed corn as the dried and brittle leaves, combined with the lower content of grain, might increase porosity of the silage. Under these scenarios, the use of inoculants to enhance fermentation is highly advised.

8. Conclusions

Abiotic stresses such as drought and heat stress can substantially affect corn silage yield and quality, although the mechanisms by which they act are different. Depending on the moment at which occurs, drought stress can have varying impacts. If drought stress occurs only at vegetative stages, dry matter yields can be compromised but not necessarily its nutritional composition. Alternatively, if drought stress occurs during reproductive stages (i.e., silking), both dry matter yield and nutritional composition can be affected. Heat stress, defined as temperatures above 35 °C, during the initial stages of kernel development can have a major negative impact in both corn silage yields and nutritional composition. Management practices, such as hybrid selection and planting date, should be considered to avoid silking and early kernel development during season of very high environmental temperatures.

Acknowledgements

The authors thank Ms. Christy L. Teets for her assistance in revising this manuscript. The authors also acknowledge the partial support of this project by USDA-NIFA Hatch Project VA-160025 and USDA-NIFA Multistate Project VA-136291 (NC-2042, Management Systems to Improve the Economic and Environmental Sustainability of Dairy Enterprises).

Author details

Gonzalo Ferreira* and Alston N. Brown

*Address all correspondence to: gonf@vt.edu

Department of Dairy Science, Virginia Tech, Blacksburg, VA, United States

References

[1] USDA, Cattle. 2014, Washington, DC: National Agricultural Statistics Service.

[2] USDA, Crop Production 2013. 2014, Washington, DC: National Agricultural Statistics Service.

[3] NRC, Nutrient Requirements of Dairy Cattle: Seventh Revised Edition, 2001. 2001, Washington, DC: The National Academies Press. 408.

[4] Ferreira, G., et al., The interaction of drought stress and heat stress as determinant of dry matter yield and nutritional composition of maize (Zea mays L) whole-plant for silage. Maydica, 2015. 6060(1):M6.

[5] Boomsma, C.R., et al., Maize morphophysiological responses to intense crowding and low nitrogen availability: an analysis and review. Agronomy Journal, 2009. 101(6): p. 1426-1452.

[6] Watson, S.A., Structure and composition, in Corn: Chemistry and Technology, S.A. Watson and P.E. Ramstad, Editors. 1987, St. Paul, MN: American Association of Cereal Chemists, p. 52-82.

[7] Abendroth, L.J., et al., Corn Growth and Development. 2011, Ames, Iowa: Iowa State University Extension.

[8] Wu, X., A. Skirpan, and P. McSteen, Suppressor of sessile spikelets1 functions in the ramosa pathway controlling meristem determinacy in maize. Plant Physiology, 2009. 149(1): p. 205-219.

[9] Oury, V., F. Tardieu, and O. Turc, Ovary apical abortion under water deficit is caused by changes in sequential development of ovaries and in silk growth rate in maize. Plant Physiology, 2015. doi: http://dx.doi.org/10.1104/pp.15.00268

[10] Çakir, R., Effect of water stress at different development stages on vegetative and reproductive growth of corn. Field Crops Research, 2004. 89(1): p. 1-16.

[11] NeSmith, D.S. and J.T. Ritchie, Effects of soil water-deficits during tassel emergence on development and yield component of maize (Zea mays). Field Crops Research, 1992. 28(3): p. 251-256.

[12] Borrás, L., et al., Coupling time to silking with plant growth rate in maize. Field Crops Research, 2007. 102(1): p. 73-85.

[13] Schoper, J.B., et al., Plant factors controlling seed set in maize: the influence of silk, pollen, and ear-leaf water status and tassel heat treatment at pollination. Plant Physiology, 1987. 83(1): p. 121-125.

[14] Hanft, J.M. and R.J. Jones, Kernel abortion in maize: I. Carbohydrate concentration patterns and acid invertase activity of maize kernels induced to abort in vitro. Plant Physiology, 1986. 81(2): p. 503-510.

[15] Cheikh, N. and R.J. Jones, Disruption of maize kernel growth and development by heat stress (role of cytokinin/abscisic acid balance). Plant Physiology, 1994. 106(1): p. 45-51.

[16] Roth, G.W., Weather effects on corn silage, in Field Crop News [Volume 1(12)]. 2001, Penn State Extension, University Park, PA.

[17] Vincent, D., et al., Water deficits affect caffeate O-methyltransferase, lignification, and related enzymes in maize leaves. A proteomic investigation. Plant Physiology, 2005. 137(3): p. 949-960.

[18] Alvarez, S., et al., Metabolomic and proteomic changes in the xylem sap of maize under drought. Plant, Cell & Environment, 2008. 31(3): p. 325-340.

[19] Farooq, M., et al., Plant drought stress: effects, mechanisms and management, in Sustainable Agriculture, E. Lichtfouse, et al., Editors. 2009, Netherlands, Dordrecht: Springer, p. 153-188.

[20] Muck, R.E., L.E. Moser, and R.E. Pitt, Postharvest factors affecting ensiling, in Silage Science and Technology, D.R. Buxton, R.E. Muck, and J.H. Harrison, Editors. 2003, Madison, WI: American Society of Agronomy, Crop Science Society of America, Soil Science Society of America, p. 251-304.

[21] USDA, Crop Production 2012. 2013, Washington, DC: National Agricultural Statistics Service.

Chapter 3

Advances in Silage Sealing

Thiago F. Bernardes

Additional information is available at the end of the chapter

http://dx.doi.org/10.5772/65445

Abstract

Spoiled silage at the top and shoulders of a horizontal silo is common because of their lower density and higher aeration. Thus, avoiding or reducing aerobic deterioration in the peripheral areas of the silages becomes a key factor for commercial farms. There are two factors that affect the top spoilage: the quality of the plastic film and how well it is held to the forage. The quality of the plastic film is related to oxygen permeability, thickness, and ultraviolet blocking. To hold the sheet to the crop, sidewall plastic associated to gravel bags and used tires have been good alternatives to be used as weights to secure the sheet on the top surface, but many other means can be applied like sidewall disks. Preventing silage losses due to an inappropriate sealing is important, both from nutritional and economic contexts. Proper air sealing produces well-fermented silage and mitigates losses in the upper layer of the silo.

Keywords: plastic cover, aerobic deterioration, dry matter losses, oxygen barrier film, silage storage

1. Introduction

Limiting losses in the upper silage layer is crucial for ensiling process. When no seal is applied, or when the seal is inadequate, air and moisture enter into the silo, affecting the quality of silage; therefore, silage is covered for two primary reasons. The first is to exclude rainfall because precipitation washes organic acids and other soluble feed components from the forage, and the second is to reduce exposure to air.

Oxygen enables various aerobic spoilage microorganisms to become active and to multiply themselves, resulting in aerobic deterioration [1] and substantial economic losses. The deterioration of the silage is indicated by temperature and pH increase, dry matter (DM) and nutrient losses, surface mold growth, and feed refusal by the animals.

Livestock farms can store silage in various ways such as horizontal silos (bunker and stacks), tower silos, bagged silos, or wrapped bales. Several farms prefer horizontal silos due to relatively low construction costs, greater safety compared to tower silos and high work rates for filling and unloading [2]. Nevertheless, their design allows large areas of the ensiled material to be exposed to the environment and prone to spoilage, especially in the upper layer and near the walls [3].

In horizontal silos, during the storage period, a spoiled layer is formed below the sealing sheet, known as "surface waste." Although there is also some evidence that invisible oxidation losses occur throughout the whole mass of silage during the storage period. A large percentage of the silage mass (about 25%) can be within the top 1 m depending on silo size and depth.

The most common material used to seal horizontal silos is the plastic film. The principal function of the film is to seal the forage and allow anaerobic conditions to establish [4]. Plastic films of 150–200 μm thickness have been used for this purpose. Although polyethylene (PE) sheeting has been the most common method used to protect silage near the surface, the protection provided is highly variable and often changes during storage [5]. Thus, the effectiveness of covering methods is very important to limit aerobic deterioration and losses in the large mass being protected.

This chapter presents the main factors related to sealing methods that affect the extent of aerobic deterioration in horizontal silos. Furthermore, the chapter review aims to identify proper management strategies to improve silage quality on commercial farms.

2. Unsealed silos

Along with proper harvesting and filling techniques, it is also equally important to properly cover a bunker silo. Previous studies have demonstrated that the quality and recovery of silage are compromised if horizontal silos are not covered with plastic film.

A study summarized the DM and nutrient losses when bunker and stack silos are not sealed [6]. From 1990 to 1993, the top 0.90 m of silage from 127 horizontal silos was sampled at three sites throughout the silo face. Sampling depths from the surface were 0–0.45 m (depth 1) and 0.45–0.90 m (depth 2). The silos were sealed with a single PE film of black or white-on-black (from 100 to 150 μm thick) secured with tires, sidewall disks or soil.

Losses were higher in bunkers and stacks that were not sealed. Silage locates in the peripheral area of the unsealed silos showed pH values ranging from 4.75 to 8.55, which were typical of spoiled silage. When a plastic film was applied, the organic matter losses in the upper layer (top 0.45 m) were reduced. Silage sealing also reduced spoilage losses in the second 0.45 m.

The aerobic deterioration is initially limited to the top 15–30 cm in an uncovered silo. The reason for this is that aerobic microbial activity is great enough in the upper layer to remove all of the oxygen entering into the crop either by diffusion or by convection. As the readily degradable components of the crop in the top layer are exhausted, the rate of microbial activity declines allowing oxygen to move deeper in the silo and cause deterioration at that level [7].

Economic evaluations indicate that the reduced losses from using a cover return more than $8.00 for each $1.00 invested in plastic and labor to cover a bunker silo [8]. In a 200-t bunker silo (6 m wide by 20 m long by 2.5 m deep), an effective seal to protect the top 1 m of silage can prevent the loss of 100–400 dollars worth silage, depending on the value of the crop. Proper sealing with a plastic cover is therefore essential to reduce losses and prevent microbial deterioration, which may result in the presence of toxins.

3. Lining bunker walls with plastic

A large part of the silage stored in horizontal silos is exposed to air and is prone to spoilage, especially in the upper part near the walls (at the shoulders of the silo), which are difficult to seal properly. A research reported silage DM losses near the surface of bunker silos to be the highest (76%) near the silo wall and the lowest (16%) in the core [9]. Thus, a problem still not fully solved is the connection of the cover to the bunker silos.

The best results are achieved by putting an additional film 1–2 m deep (depending of the silo size) between wall and forage, and then over the forage, before the main sheet is attached (**Figure 1**). The result of this additional effort is that silage quality along the wall is similar as that throughout the silo [10, 16].

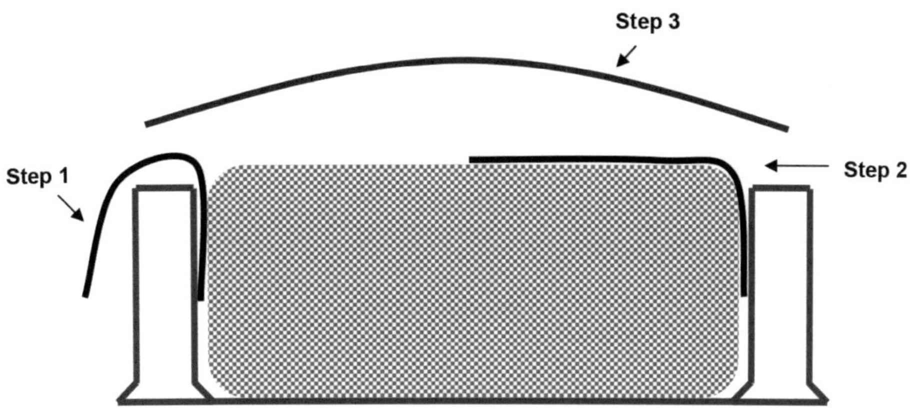

Figure 1. Bunker lining diagram. Step 1 = before silo filling, place a plastic sheet along the length of the sidewall with approximately 2 m of excess draped over the wall; Step 2 = sidewall plastic should lap onto the forage top at the end of filling; and Step 3 = cover the bunker with additional plastic film.

There are limited studies showing the effects of bunker silo sidewall plastic on silage characteristics. A survey in 20 dairy farm bunker silos, 10 without and 10 with sidewall PE plastic, demonstrated that lining bunker wall improves fermentation and produces silage with greater digestibility [11]. Sidewall plastics have more effects on forage preservation; however, it will be addressed in Section 4.2 of this chapter.

4. Plastic film to cover silage

A plastic film to cover silage has to fulfill three essential functions. First, the film should prevent precipitation and damage caused by meteorological effects and animal attack. Second, the film should be UV resistant to resist prolonged exposure to sunlight. Finally, the third function of the silo film is guarantee anaerobic conditions in the silage.

4.1. Color and thickness of plastic film

The color of sheet should affect the amount of air infiltration and subsequent aerobic losses because oxygen permeability into the silage is highly dependent on the temperature of the plastic. Only few data have been published about the thermal effects of covers on the upper silage layers. It is important to emphasize that these surface layers are highly susceptible to poor fermentation because of unsatisfactory packing density and the proximity to the plastic film. Moreover, a microclimate in the upper layer created by the high temperature influences strongly the growth of undesirable microorganisms (yeasts, molds, and aerobic bacteria).

This is consistent with the observations by Bernardes et al. [12], who found highest DM losses and yeast counts when corn silages were sealing with black PE. Black sheet also shows higher temperature in relation to white-on-black film during storage period (**Figure 2**).

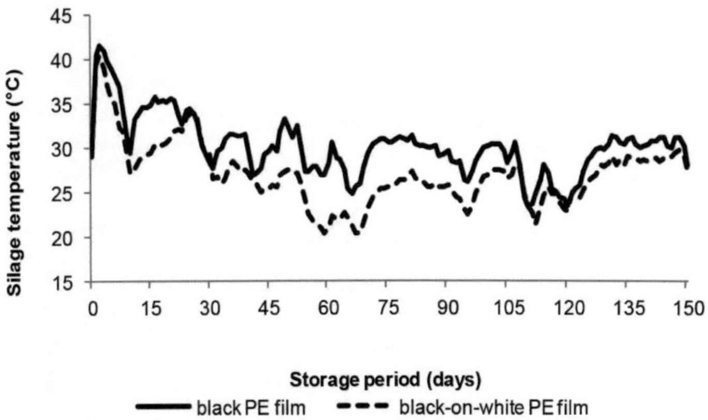

Figure 2. Effects of the color of plastic film on temperature of corn silages during 150 d of storage.

A study reported the effects of the color on the temperature of the film surfaces [13]. The authors found that in the morning hours, temperature peaks were up to 16°C higher for the black film in comparison with the white film. As expected, the highest values were reached at midday, with the black and green colored films showing a very similar thermal behavior. The same applied for the evening hours.

A model to establish the costs of plastic and respiration losses because of air penetration through the film was developed by Savoie [5]. To calculate the optimal thickness, the following parameters were considered: storage period, silage density and DM content, film permeability, and the relative value of plastic and silage. Polyethylene silage bags of different thickness (100, 150, and 200 μm) did not produce significant differences in losses in 130 d, averaging 0.2% loss/month when perfectly sealed [5]. However, modeling of different film thickness indicated that 100 μm was economically optimum on a stack silo for 3 months storage, 150 μm for 7 months, and 200 μm for 12 months. It is important to emphasize that films with thicker thickness have more puncture and tear resistance than the thin ones.

4.2. Oxygen permeability of plastic films

Air is the major cause of spoilage in silage. Polyethylene is not totally impermeable to oxygen diffusion and thus will not completely prevent oxygen ingress. There is a general agreement, therefore, that low oxygen permeability of the sheets has to be sought.

The first generation of barrier films emerged in the early 2000s when a co-extruded PE-polyamide film was developed for covering horizontal silos [14]. It had 125 μm in thickness and comprised two outer layers of PE with a central layer of polyamide. However, this film showed some problems such as rigidity and fragility what led to less use in farm conditions.

More recently, oxygen barrier (OB) films made with PE and ethylene-vinyl alcohol (EVOH) have been available. Ethylene-vinyl alcohol combines the highest barrier properties with good mechanical characteristics such as puncture resistance, tear resistance, and stretch properties [15].

There are two types of OB films, which are available on European and American market, respectively. The first one is a white-on-black sheet, which is composed by a layer of EVOH between layers of PE during the manufacturing process. The second is a thin film (45-μm-thick PE + EVOH), which needs to be covered by tarp or a second layer of PE during its application in practical conditions. This procedure is necessary because it is not UV stabilized. Originally, the thin OB film was associated with a tarp to protect from UV light as well as from physical damage. However, this type of UV cover is expensive for producers with modest resource availability. Thus, to overcome this problem, a method that combines the thin film with a conventional PE sheet has been created. An experiment was carried out to evaluate the effectiveness of this method for covering corn silage in bunker silos [16]. Two systems were assessed, as follows: the first method comprised a sheet of 45-μm-thick OB film placed along the length of the sidewall before filling, with approximately 2 m of excess draped over the wall. After filling, the excess film was pulled over the wall, and a sheet of PE was placed on top. The second system involved using a standard sheet of 180-μm-thick PE film. Over 2 years, eight commercial bunker silos were divided into two parts lengthwise so that half of the silo was covered with OB and the other with standard system. Oxygen barrier method produced well-fermented silages, which were similar to the central part of the silo (core), whereas PE system showed less lactic acid and greater pH and mold counts compared with core. The estimated milk yield for PE system was 116 kg/ton less than core, as OB system and core were similar (1258 and 1294 kg/ton, respectively), as shown in **Figure 3**. These results and those obtained by Borreani and Tabacco [17] showed a net economic gain when the OB films are used due to both reduced nutrient losses and labor time required to clean the upper layer, even though these films cost more than the PE layer.

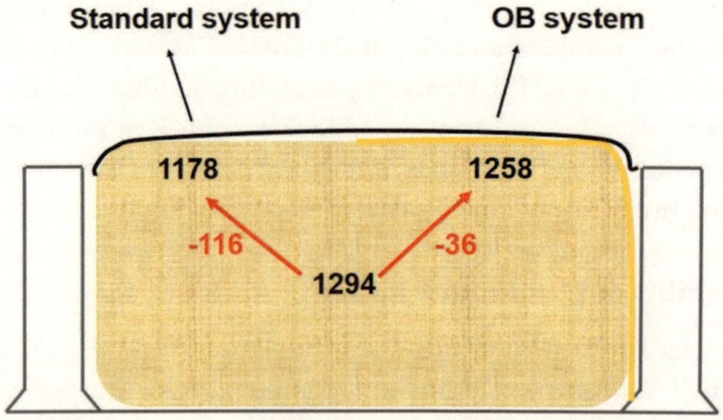

Figure 3. Effects of two covering system on estimated milk yield (kg/ton) of corn silages. Standard system = a single sheet of polyethylene (PE) film; OB system = oxygen barrier film between the silo wall and forage and covered by a second layer of PE film. *Source*: Lima et al. [16].

4.3. Biofilms

An environmental objective is to reduce the quantity of plastic used in agriculture, and there may be opportunity for achieving this by reducing the use of the plastic film for sealing silos. However, horizontal silos produce less plastic wastes than most other systems that use PE film for air tightness. Round bale silage requires at least 5.5 kg of plastic/ton DM. Stack silos use about 1.3 kg of plastic/ton DM, four times less than the round bale silage system [5].

A study was conducted to determine whether the PE film could be replaced with bio-based biodegradable films [18]. A standard 120-μm-thick white-on-black PE film and two different 120-μm-thick biodegradable plastic films were used to produce the silage bags for that experiment. The results of this research showed that the development of new degradable materials to cover silage could be possible. In addition, the authors recommended that further research should be undertaken to improve the blend for enhancing film stability over time and its resistance under outdoor conditions.

5. Weighting the plastic cover

To prevent deterioration in horizontal silos, the common practice is to use plastic film held in place with used car tires. Tires have been widely used because of their low cost and ready availability. In a study reported by Ruppel [19], there was a reduction in the temperature and improved protein availability of hay crop silage when the number of tires per square meter increased. The effects of several covering methods on reduction in the silage losses in the top layer concluded that higher tire density (30 tires per 10 m^2) and sand bags along the shoulders resulted in lower losses [19].

The results of a study on different silage sealing systems were presented by Borreani and Tabacco [20]. A farm bunker silo was covered with a single white-on-black sheet. Half of the width of the sheet was covered with tires (25 kg/m^2), and the other half was covered with gravel (200 kg/m^2). The silo was opened for summer consumption and had a low feed-out rate (12 cm/d). The results showed that the difference in sealing system affected the temperature in the peripheral areas of the corn silage. The silage covered with tires reached a maximum temperature exceeding 40°C, whereas that covered with gravel did not.

The amount of soil placed on top of the PE plastic cover also has an effect on silage quality. The effectiveness of several sealing strategies that are used in Brazil on reduction in losses in the top layer was tested by Griswold et al. [11]. Covering a black plastic sheet with soil (100 kg/m^2) reduced losses, and this was associated with decreased pH and ash content and lower counts of yeasts. However, most farmers are very reluctant to cover horizontal silos with soil, particularly if the silo is large because they do not believe that the labor and costs involved in covering with soil are reasonable and economical. Moreover, the soil used as a cover can contaminate the silage during unloading. Thus, alternative covering strategies to reduce aerobic deterioration in the peripheral areas of the corn silage in a warm climate were investigated. Three treatments were evaluated: (1) black PE film (control); (2) black PE film plus sugarcane bagasse (10 kg/m^2) over the sheet; and (3) black PE film plus soil (30 kg/m^2) over the sheet [21]. Treatments did not affect the temperatures during the early part of the storage period, but after about 80 d of fermentation, the temperature started to rise in the control silage but not in the others. This can be attributed to the effect of oxygen permeability of the film during a long storage period because the gas transmission rate is reduced by the presence of soil or sugarcane bagasse over the sheet. These results also suggest that the material over the film reduces billowing caused by the wind what affects the amount of air drawn into the silo.

It is important to emphasize that keeping the plastic cover weighed down is critical during the storage and feed-out periods. During the unloading, air can penetrate the peripheral areas of a silo up to 1 m or more beyond the feed-out face [10], especially when the sealing cover is not weighed down or is weighed only with tires, suggesting that, in these situations, daily removal rates should be higher than 30 cm/d to avoid extended aerobic spoilage.

6. Chemical additives on the top of the silos

Especially in warm climates, whole-crop cereal silages such as corn, sorghum, and wheat are susceptible to aerobic deterioration. This is because aerobic yeasts are most active at 20–30°C [22]. Therefore, efforts need to be made to protect the silage near the surface when PE films are used. A research evaluated the application of additives (sodium benzoate and *Lactobacillus buchneri*) directly to the top of the silage and concluded that sodium benzoate applied at a 2 g/kg rate was the most suitable additive to improve the fermentation, reduce the aerobic deterioration, and preserve the nutrients of corn silage at the top of bunker silos [23]. Results from this study showed that the *in vitro* digestibility of the silage at the core and those treated with sodium benzoate were above 640 g/kg, whereas silages untreated and treated with two strains of *L. buchneri* had values close to 600 g/kg (**Figure 4**). According to the authors, under field conditions, the strains may have had their growth affected by high temperatures, and thus, chemical additives present more robust effects than biological ones when applied at the top.

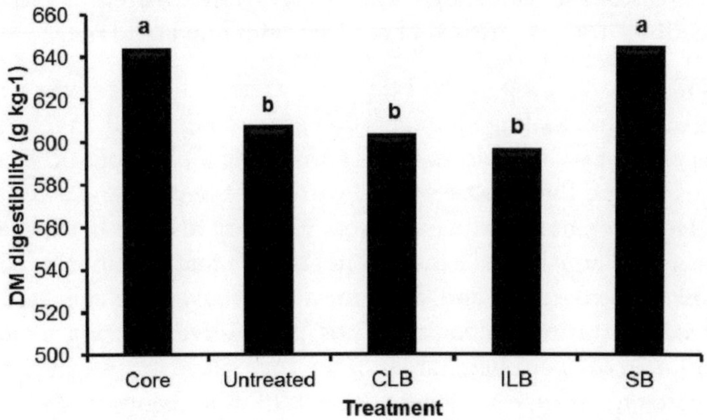

Figure 4. Effects of additives on *in vitro* DM digestibility in different zones of the bunker corn silage. Core = silage in the core of the silo; CLB = silage treated with commercial *Lactobacillus buchneri*; ILB = silage treated with indigenous *L. buchneri*; SB = silage treated with sodium benzoate. *Source*: Da Silva et al. [24].

7. Conclusions

The detrimental effect of air at silage near the surface is a key point to avoid losses of dry matter and quality. To date, no alternative to the use of plastic in covering bunkers or stacks has proven commercially viable for silage producers. Given the widespread use of horizontal silos worldwide, it is vitally important that the film used possesses good oxygen barrier properties as well as good mechanical properties.

In horizontal silos, the plastic needs to be held tightly to the crop. This is usually accomplished with used tires, but many other means can be applied. Besides that, lining bunker walls with plastic improve silage quality along the walls.

The silos' sealing will continue evolving to meet future needs in a conservation of fresh forage, minimize loss and cost, reduce environment contamination, and provide a safe and efficient on-farm feeding system.

Author details

Thiago F. Bernardes

Address all correspondence to: thiagobernardes@dzo.ufla.br

Department of Animal Science, Federal University of Lavras, Lavras, Minas Gerais, Brazil

References

[1] Pahlow, G., R. E. Muck, F. Driehuis, S. J. W. H. Oude Elferink, and S. F. Spoelstra. 2003. Microbiology of ensiling. In D. R. Buxton, R. E. Muck, and J. H. Harrison, eds. Silage Science and Technology. Vol. 42. Madison, WI, Am. Soc. Agron., Crop Sci. Soc. Am., and Soil Sci. Soc. Am. pp. 31–93.

[2] Savoie, P. and J. C. Jofriet. 2003. Silage storage. In D. R. Buxton, R. E. Muck, and J. H. Harrison, eds. Silage Science and Technology. Vol. 42. Madison, WI, Am. Soc. Agron., Crop Sci. Soc. Am., and Soil Sci. Soc. Am. pp. 405–465.

[3] Bolsen, K. K., J. T. Dickerson, B. E. Brent, R. N. Sonon Jr., B. S. Dalke, C. Lin, and J. E. Boyer Jr. 1993. Rate and extent of top spoilage losses in horizontal silos. J. Dairy Sci. 76:2940–2962.

[4] Bernardes, T. F., L. G. Nussio, and R. C. do Amaral. 2012. Top spoilage losses in maize silage sealed with plastic films with different permeabilities to oxygen. Grass Forage Sci. 67:34–42.

[5] Savoie, P. 1988. Optimization of plastic covers for stack silos. J. Agric. Eng. Res. 41:65–73.

[6] Berger, L. L., and K. K. Bolsen. 2006. Sealing Strategies for Bunker Silos and Drive-Over Piles. Online. http://www.oznet.ksu.edu/pr_silage/publications/NRAES%20Berger%20and%20Bolsen%20Sealing%20Strategies%204-14-06.pdf [Accessed 10.08.15].

[7] Muck, R. E. 1998. Influencing of Air on Ensiling. ASAE Paper 981054. ASAE, St. Joseph, MI.

[8] Rotz, C. A. and R.E. Muck. 1993. Silo selection: balancing losses and costs. In: Proceedings of Silage: Seed to Animal. NRAES Publ. 67. Ithaca, NY. pp. 134–143.

[9] Asbell, G., and Y. Kashanci. 1987. Silo losses from wheat ensiled in bunker silos in a subtropical climate. J. Sci. Food Agric. 40:95–98.

[10] Honig, H. 1991. Reducing Losses During Storage and Unloading of Silage. Silage Conservation Towards 2000, Braunschweig, Germany. G. Pahlow and H. Honig, ed. Institute of Grassland and Forage Research and Federal Research Center of Agriculture Braunschweig-Völkenrode, Braunschweig, Germany.

[11] Griswold, K. E., E. E. McDonell, L. Kung, Jr., and P. H. Craig. 2010. Effect of bunker silo sidewall plastic on fermentation, nutrient content and digestibility of corn silage. J. Anim. Sci. 88 (E-Suppl.2):622.

[12] Bernardes, T. F., L. G. Nussio, R. C. Amaral, and A.L.B. Schogor. 2009. Sealing strategies to control the top losses of corn silage. In Proc. 15th Int. Silage Conf., Madison, Wisconsin, USA pp. 213–214.

[13] Snell H. G. J., Oberndorfer C., Lucke W. and Van Den Weghe H. F. A. 2003. Effects of polyethylene colour and thickness on grass silage quality. Grass Forage Sci. 58: 239–248.

[14] Borreani, G., E. Tabacco, and L. Cavallarin. 2007. A new oxygen barrier film reduces aerobic deterioration in farm-scale corn silage. J. Dairy Sci. 90:4701–4706.

[15] Borreani, G., E. Tabacco, and D. Deangelis. 2011. Special EVOH based films improve quality and sanity of farm corn silage. In Proc. Agricultural Film 2011: Int. Conf. Agricultural and Horticultural Film Industry. Applied Market Information Ltd., Barcelona, Spain. pp. 1–15.

[16] Lima, L. M., Dos Santos, J. P., De Oliveira, I. L., Gusmão, J. O., Bastos, M. S., Da Silva, S. M. and T. F. Bernardes. 2016. Lining bunker wall with oxygen barrier film reduces nutrient losses of corn silages. J. Anim. Sci., 94:312.

[17] Borreani, G. and E. Tabacco. 2014. Improving corn silage quality in the top layer of farm bunker silos through the use of a next-generation barrier film with high impermeability to oxygen. J. Dairy Sci. 97:2415–2426.

[18] Borreani, G. and E. Tabacco. 2014. Bio-based biodegradable film to replace the standard polyethylene cover for silage conservation. J. Dairy Sci. 98:386–394.

[19] Ruppel, K. A. 1993. Bunker silo management and its factors on haycrop quality. In Silage production from seed animal. NRAES-67. Syracuse, NY. 23–25 Feb. 1993. Northeast Reg. Agric. Eng. Ser., Ithaca, NY. pp. 67–84.

[20] Ashbell, G., and Z. G. Weinberg. 1992. Top silage losses in horizontal silos. Can. J. Eng. 34:171–175.

[21] Borreani, G. and E. Tabacco. 2007. Il Silomais: Guida Pratica. Quaderni della Regione Piemonte – Agricoltura. Torino, Italy.

[22] Amaral R. C., Queiroz B. C. Garcia E. H. C., Sá Neto A., Bernardes T. F. and Nussio L.G. (2010) Aerobic deterioration in maize silages under different covering methods of the plastic film. In: Schnyder H., Isselstein J., Taube F., Auerswald K., Schellberg J., Wachendorf M., Herrmann A., Gierus M., Wrage N. and Hopkins A. (eds) Proceedings of XXIII General Meeting of the European Grassland Federation, Kiel, Germany, 2010, pp. 83.

[23] Ashbell, G., Z. G. Weinberg, Y. Hen, and I. Filya. 2002. The effects of temperature on the aerobic stability of wheat and corn silages. J. Ind. Microbiol. Biotechnol. 28:261–263.

[24] Da Silva, N. C. Dos Santos, J. P. Ávila, C. L. S., Evangelista, A. R. Casagrande, D. R. and T. F. Bernardes. 2014. Evaluation of the effects of two *Lactobacillus buchneri* strains and sodium benzoate on the characteristics of corn silage in a hot-climate environment. Grassland Sci. 60:169–177.

Section 2

Non-Conventional Crops

Chapter 4

Ensiling of Forage Crops in Semiarid Regions

João Paulo Farias Ramos, Edson Mauo Santos,
Ana Paula Maia dos Santos,
Wandrick Hauss de Souza and Juliana Silva Oliveira

Additional information is available at the end of the chapter

http://dx.doi.org/10.5772/101990

Abstract

Edaphoclimatic condition of the semiarid region is unfavorable for the forage production of livestock. Silage is considered a better alternative to conserve forage crops. Ensiling is a technique for preserving forage, in which the ensiled mass is acidified under anaerobic conditions. The lactic acid bacteria present in the environment produce lactic acid, thereby making the environment acidic, and convert soluble substrates into organic acids. Many microorganisms are involved in the fermentation process of silage and their development depends on the characteristics of ensiled materials, such as dry matter, water-soluble carbohydrate content, buffering capacity and presence of indigenous microbial. Ensiling is a favorable technque used in the semiarid region because it preserves the nutritional values of the crops and the water. Some plant species are produced in semiarid regions because they are resistant to water deficit and high solar radiation. The main crops of semiarid regions are sorghum, pearl millet, grasses, cactus pear, and leguminous. Due to agronomic conditions available for their production during periods of rain, for ensiling these plants are important for the fermentation profile of each species because the ratio of the dry matter to water-soluble carbohydrate content and buffering capacity directly influence the end product of silage.

Keywords: cultivate crop, drought, forage preservation, forage silage, tropical crop

1. Introduction

The semiarid regions are characterized by an irregular distribution of rainfall with greater variability between years and within the same year, and high solar radiation. This hydric variability originates from complex systems of the formation of rain, with the occurrence of concentrated rain in a few months of the year and alternate years, irregularity, existence of geographic barrier concentrating higher humidity in the valleys and leaving dried slopes, and variability of soil with greater or lesser ability to retain water [1].

The variability of rainfall provides the diversity of fauna and flora species in the semiarid region. The soil and climate conditions are associated with the characteristics of species, such as solar radiation, sunshine, and air temperature. The climatic variations suggested that animal production systems operate according to the availability of resources and controlled principally by the availability of water, adopting rational strategies for production.

In semiarid regions, annual rainfall is irregular, low, and highly variable in space and time, with permanently high evapotranspiration rate. Therefore, the agricultural systems used in the semiarid region should be based on plants that develop efficiently and quickly by using the resource availability of pulses [2], because the water dynamics is the main variable for controlling the transformation process of individual nutrients available for plants.

Tropical regions, such as semiarid regions of Brazilian, may have a high capacity for forage production, but climatic variables make it difficult for the development of the animal production system. The quantity and quality of forage are key factors for animal production. The management also influences the forage characteristics and animal production. During the dry season, a significant reduction in native vegetation occurs, and this affects animal forage production.

The shortage of forage during the dry season and low nutritive value of forage may compromise the animal production, resulting in decreased productivity. In this situation, the producers become dependent on the availability of the preserved forage, hay, and silage, cultivated forage crops, and crop residues to feed cattle in the semiarid region [3].

For the efficient production of forage in the semiarid regio, it is essential to know conditioning factors inherent to soil, climatic condition, and plant interaction mechanisms to drought and production capability. Adapted crops, due to their efficiency in the accumulation of green matter in these climatic conditions, are available as a more viable option to the semiarid region. Among other considerations. the forage conservation practices, silage is a better alternative to reduce the qualitative and quantitative fluctuations in the supply of forage to the animals.

As a literature review, this chapter presents scientific reports on ensilage of forage with productive potential for the semiarid regions. This study presents the main crops for ensilage in the semiarid regions and their fermentation characteristics.

2. Influence of drought on the production of the crops in the semiarid regions

Drought is the meteorological event when there is inadequate water availability in the soil or rainfall, including quantitative and qualitative, during the life cycle of a plant, limiting full expression of the gene of the plant potential and preventing the maximum yield from a culture [4].

The planning activities of animal production in drought regions should take into consideration some factors, such as production yield, drought resistance, and water-use efficiency by plants, for crop production. Although a culture presents high production yield, this may not be compatible with higher drought resistance or increased water-use efficiency [5].

In rainfed situations where crops depend on unpredictable seasonal rains, the maximum use of soil moisture is a crucial component for drought resistance, that is, water-use efficiency allows the production yield even in the situations of water deficit [5].

Drought resistance is the ability of a plant to produce with minimal loss in a water-deficit environment. Drought resistance mechanisms can be classifieds into the following three categories: drought escape, drought avoidance, and drought tolerance [6].

Drought escape is the ability of plants to complete the life cycle before there is a serious water deficit for plant and soil. The phenological development of plants is fast with early flowering and maturity and the duration of the growing season depends on water deficit [6]. The success of these species depends on the efficient reproduction before a more intense water stress. With short life cycle and high growth rates and storage, this process uses reserve for seed production [7].

Drought avoidance is the ability of plants to maintain some potential of water in the tissue even with low moisture in the soil. The better absorption of water, mobile water storage, and reduction of water losses are some of the processes that are used for these plants. The balance between turgor, increased depth of rooting, higher absorption efficiency, and lower losses of water allows the survival of plants under dry conditions [6]. Furthermore, older leaf senescence reduces the energy cost of the plant [7] directing all the energy to dry the adaptive mechanism.

Drought tolerance is the ability of plants to resist water deficit with low potential of water in the tissues [6]. In water-limited environments, the plants can produce forage mass using water maintenance mechanisms in the plant. One of the processes is the osmotic adjustment. Osmotic adjustment is an adaptive response to cellular stress that in some cultures increases the avoidance dehydration and supports the production yield under stress [5]. Osmotic adjustment maintains turgor and resists to dehydration through solute accumulation in the cell, an increase in cell elasticity, and a decrease in the cell size [6].

In drought, these plants maintain the water content accumulating several nontoxic solutes that do not interfere with metabolism; these are compatible solutes such as fructan, trehalose, polyols, glycine, proline, betaine, and polyamines [4, 8].

Although the drought resistance is important for crop production in semiarid, the adjustments resulting from this drought tolerance have disadvantages because of lower production output. The stomata closure and reduction in leaf area result in lower carbon dioxide assimilation and higher osmotic adjustments that can have a negative effect on the plant energy requirement [4].

Cultivate crops are grown using more than one mechanism to resist drought [9]. Thus, it is interesting to note that the adaptive mechanisms of crops grown in the semiarid region have a balance of escape, avoidance, and drought tolerance, maintaining the yield production as much as possible. Through conventional breeding or biotechnological methodology, the development of superior genotypes resistant to drought is possible [4].

Most of the crops produced in the world are sensitive to water deficit. Even cultivative crops, such as pearl millet, sorghum, and pigeon pea, in semiarid regions are affected by drought during the reproductive stage [4].

The C4 plants are considered to be dominant in resistance and drought tolerance because they are capable of maintaining photosynthesis with closed stomata. Even with the small reduction of photosynthesis under water stress conditions, the C4 crops such as sorghum and panicum have the ability to grow in dry region and are considered to have a great potential for enhancing forage production and food security in the world [7]. The C4 plants provide competitive conditions of low availability of water, high temperatures, and high light intensities [10]; they have high water-use efficiency and mechanisms for CO_2 concentration [7].

Other factor that influences the nature of response of plants to drought is the thermal stress. Thermal stress can reduce transpiration and can dehydrate the plant cells, reduce the availability of nutrients, and cause osmotic stress together with the drought. In the plant growth stage, water stress can interfere with the final yield production of the crop [7]. Corn yield, for example, is a culture that is extremely sensitive to water stress during the period of the previous life cycle of flowering. Crops such as sugarcane may have a greater impact of water stress when its leaves are establish than in the initial period, which may affect the final yield [7, 11].

The adaptive responses are based on complex changes to cope with stress, primarily to maintain water potential in main tissues. Crops such as sorghum and pearl millet are drought tolerant and cultivated on a large scale in the semiarid region. These crops are able to maintain photosynthetic activity under water stress conditions and thus increase the final yield [12].

The osmotic adjustment required for drought tolerance forage can increase the solute values as fructan [4, 8] increases the values of soluble carbohydrate in these forage.

The concentration of water-soluble carbohydrates (WSC) in ensiled materials influences the fermentation profile because the WSC concentrations are used for the production of lactic acid [13]. The minimum content of WSC to appropriate fermentation of good silage varies between 6 and 12% [14]. In contrast, a large amount of WSC concentration may predispose to undesirable occurrence of fermentation realized from yeasts because of the excessive lactic acid production, which leads to losses resulting from the alcoholic fermentation [15].

In the semiarid region, there is a tendency that the forage contains a higher WSC content. The forage sorghum, pearl millet, and buffel grass show a WSC concentration (DM basis) of 13–20, 9, and 3.1%, respectively [16].

3. Ensiling process

Ensiling is a method of forage conservation. It is based on natural fermentation, in which lactic bacteria convert the WSC into organic acids (principally lactic acid) under anaerobic conditions. As a result pH decrease and the silage is preserve [17]. The primordial objective of forage ensiling is to preserve the original composition of nutrients found *in natural* plant during storage with minimum losses [18].

The forage conservation as silage depend on favorable conditions, such as the amount sufficient WSC to lactic acid production and low buffering capacity, which promote rapid lowering of pH that inhibits the growth of some deleterious microorganisms, maintaining the nutritional values of forage.

Before the ensiling process, aerobic and facultative anaerobic microorganisms are able to grow in high pH and predominance. As long as pH decrease and oxygen is consumed, the anaerobic and anaerobic facultative acid tolerant bacteria grow in the environment.

Ensiling is divided into four phases with different time and intensity [19, 20].

- Aerobic phase: It occurs during filling of silo and extends until a few hours after the packing of silo. The aerobic phase is undesirable because all obligatory and facultative aerobic microorganisms (yeasts, molds, and bacteria) are active in this phase, but it is an inevitable phase. As it is associated with the fermentable substrate and energy losses, it is important to reduce the duration of this phase. It recommended that the forage be chopped, compacted, and rapid packing of the silo [13]. The final stage of the phase includes exhaustion of oxygen in silo.

- Active fermentation phase: After exhaustion of oxygen in silo, there is a decrease in silage pH because of organic acids production from WSC. In initial, enterobacteria and heterofermentative lactic bacteria grow in ensiled mass. With the larger decline of pH, homofermentative lactic bacteria dominate the anaerobic environment. In this phase, there is the more production of organic acid, such as acetic and lactic acids, and also ethanol and CO_2. The major growth of lactic acid bacteria (LAB), and consequently, larger lactic acid formation inhibit the development of other microorganisms, principally due to lowering of pH. This phase extends to the stability and reduces excessive microbial activity.

- Stability phase: It a phase with low biologic activity, since it does not penetrate air in the ensiled mass. The pH permanence is stable in 3.8–4.2, inhibiting microbial activity. Only some acid tolerant enzymes maintain activity [20]. The acid pH and anaerobic conditions maintain the ensiled mass stability to the silo opening.

- Discharge phase: It occurs at the opening of the silo and expose the ensiled mass to high oxygen concentration, which favors the growth of enterobacteria, molds, yeasts, and other microorganisms. Yeasts are the first microorganism to develop in silage after the opening, causing deterioration of the conserved forage [13]. There are heat and CO_2 production due respiration, which results the decrease in lactic acid and residual WSC, and increase in silage pH [13]. The appropriate management may minimize the losses after opening of silo.

3.1. Microorganism involved in the ensiling process

The ensiling process is complex and variable. It consists, basically, in conjunct action of the large number of microorganisms and may be considered a metabiose because it occurs at simultaneous and successive development of different microorganisms that depends on specific pH, substrates, and potential redox in environment of silo.

The microorganisms present in plant before ensiling may be aerobic and anaerobic, desirable and undesirable to fermentation. **Table 1** presents the most common types of microorganisms and their presence in plants.

Groups	pH
Total aerobic bacteria	>10,000,000
Lactic acid bacteria	10–1,000,000
Enterobacteria	1000–1,000,000
Yeasts	1000–100,000
Molds	1000–10,000
Clostridia	100–1000
Bacillus	100–1000
Acetic acid–producing bacteria	100–1000
Propionic acid–producing bacteria	10–1000

Source: Adapted from Pahlow et al. [20].

Table 1. Typical bacterial and fungal population of plants groups before ensiling.

				Recuperation (%)	
Organism	Rota	Substrate	Product	Energy	DM
LAB	Homofermentative	Glucose	2 Lactate	96.9	100
LAB	Heterofermentative	Glucose	1 Lactate + 1 Acetate + CO_2	79.6	83
LAB	Heterofermentative	Glucose	1 Lactate + 1 Ethanol + CO_2	97.2	83
Yeast		Glucose	2 Ethanol + 2 CO_2	97.4	51
Clostridia		Glucose	1 Butyrate + 2 CO_2	77.9	66
Enterobacteria		2 Glucose	1 Lactate + 1 Acetate(1 Ethanol) + CO_2	88.9	83

Source: Adapted from de McDonald et al. [13].

Table 2. Acidify efficiency and fermentation and main fermentative routes of microorganisms in silage.

The microorganisms present in plants are diverse in genera and species with different fermentative routes. Each group has specific temperature and substrate to grow with higher or lower energy demand. In the fermentation process, microorganisms convert soluble substrates into organic compounds. **Table 2** presents the main fermentative routes of microorganism in silage.

The growth of lactic acid bacteria in ensiled mass is important because its metabolism does not result in considerable DM losses, following the principle of forage preservation. The LAB converts one mole of glucose to two moles of lactic acid without DM losses [13].

In situations where the forage has a low amount of substrates may have predominant of other microorganisms, such as enterobacteria, because the pH is not low sufficiently. In the opposite situation, ensiling of forage with excess WSC may be in the presence of acid tolerant microorganisms such as yeasts that are able to consume lactic acid and WSC. The excess WSC in the plant leads to the formation of acid silage due to excessive lactic acid production. Acid silage, such as sugarcane and saccharine sorghum silages, has high ethanol concentration because of alcoholic fermentation.

3.2. Characteristics of forage to ensiling

The dry matter content is an important factor that affects the fermentation and preservation of ensiled mass. The ideal DM content is between 30 and 35% [13]. However, research studies indicated the values to corn silage being necessary attempt to characteristics of each culture, because it might occur good fermentative profile in silage of forage with inferior DM values.

Generally, the high content of moisture favors undesirable microorganisms, such as *Clostridium* and enterobacteria that are butyric and acetic acid and ammonia producers, implying in nutrient losses. However, higher DM content impacts the compaction and reduction of air present [21].

The WSC concentration in ensiled materials influences the fermentative profile, because the WSC concentrations are used for the production of lactic acid [13]. The minimum content of WSC required for appropriate fermentation of good silage varies between 6 and 12% [14].

Other important factors that influence the silage are the buffering capacity that is resistance to lowering the pH of ensiled materials. The compounds able of buffering the environment in forage are some organics acid, potassium, and calcium inorganic bases and nitrogen substances, such as protein and products of their degradation, free amino acids, amine, and ammonia [22]. The action of buffering capacity in silage is associated with other factors such as WSC and DM concentration. Thus, the pH of silage is determined by relationship of protein and water-soluble carbohydrate [22].

3.3. Fermentation loss in the ensiling process

The ensiling process changes the natural structure of forage and may cause some losses. Besides the natural physical losses, such as crop losses, chemical losses also occur and may compromise the energy and nutritive value of silage. Although some losses may be

Process	Losses (%)	Causative agents	Classification
Respiration	1–2	Plant enzymes	Inevitable
Fermentation	2–4	Microorganisms	Inevitable
Effluent	5–7	Moisture	Inevitable
Secondary fermentation	0–5	Plant, moisture, silo environment	Preventable
Aerobic deterioration in storage	0–10	Ensiling time, density, plant, packing	Preventable
Aerobic deterioration in discharge	0–15	Moisture, season, density, technical	Preventable
Total losses	8–43		

Source: Adapted from McDonald et al. [13].

Table 3. Losses in the ensiling process.

unavoidable, such as biochemist changes, plant respiration, and fermentation (**Table 3**), other types of losses can be avoided with appropriate practice of the ensiling procedures.

The energy and dry matter disappearance is an indicative of losses in the ensiling process. The residual respiration during filling the silo and immediately after sealing, types of fermentation, effluent production, undesirable fermentation during the storage, and aerobic deterioration are the main causes of energy and dry matter losses [21].

The losses related to respiration usually occur early. The respiration in silo initially occurs due to the presence of oxygen in the ensiled materials, thus the cellular respiration use the air oxygen and substrates producing CO_2, heat, and H_2O. Some factors can affect the respiration rate in the silo, such as temperature, which increase the initial rate of reaction and destruction of enzymes, usually by denaturation; oxygen concentration, the high amount of oxygen in the silo promotes an increase in the respiration rate and higher the temperature, and consume more energy; WSC content: the amount of soluble substrates in ensiled materials can influence the respiration, since they are consumed during respiration.

Silage fermentation usually causes DM losses due to the activity of microbial and enzymes. The losses related to the fermentation represent the highest percentage of losses in the silage process. These losses can be resulting from the production of water, gas, heat, and effluents during the fermentation process [22].

The effluent losses are associated with the DM content of plant, the activity of the water metabolism and the physical procedure of cutting and application of additives in ensiled forage [23] and DM losses can be highly variable [16]. After evaluating the sorghum silages in Brazilian semiarid we observed a variation of 10–24% DM losses.

In ensiling, besides DM losses, nutritional losses should also be taken into account. Sugarcane and sorghum silages can show high nutritional losses because of a high content of WSC, which may result in increase in alcoholic fermentation. Many studies indicate that the application of additives in the ensiled material considerably reduced these losses of substrates [24–26].

Other fermentation can also occur and reduce the nutritive value of silage, as proteolysis. The proteolysis is associated with DM, protein, WSC content, pH, and ensiling time [27]. It is an undesired reaction because the resulting products of the process (ammonia and amines, principally) indicate high nutritional losses.

In discharge of silo to offer silage to animals aerobic deterioration can also occur, which is one of the main problems after exposure to air [28]. This process occurs due the penetration of air in ensiled materials, which is favorable for the grown of aerobic microorganisms, acid tolerant, and the oxide products resulting in silage fermentative process [29]. The air exposure of silage can chance its chemical compositions and alter the nutritional value.

4. Crops for silage production in semiarid regions

The mains characteristics that determine the fermentation profile during ensiling involve the interaction of factors such as: DM content, WSC concentration, and buffering capacity of plant.

In the case of semiarid, plant species resistance to hydric deficit and climatic conditions are indicated to ensiling. The main forages are sorghum, pearl millet, tropical grasses, leguminous, and cactus pear.

4.1. Sorghum

Sorghum (*Sorghum bicolor* L. Moench) is an appropriate grass for silage with agronomic and nutritional characteristics, because it is tolerant to drought and responds even in soils with limited nutrients [30] and its phenotypic characteristics facilitate planting, management, harvesting, and storage. The other significant characteristic of sorghum is that it will regrowth after each harvesting [31].

The sorghum is a resistant to hydric deficit in semiarid. Their resistance is associated with the physiology characteristics and efficiency of rain. Researchers evaluated the efficiency of rain in sorghum genotypes in semiarid and found positive results, values between 944.37 and 126.25 kg DM/ha/mm that indicated high efficiency in covert water of rain in production [32].

In addition to their agronomic traits, sorghum has desirable characteristics for fermentation, such as a suitable dry matter content, high carbohydrate concentration, and low of buffering substance content [33, 34].

Sorghum is a crop that has desirable characteristics for the production of silage; however, as the WSC concentration is higher in the stem, forage sorghum and saccharine sorghum usually have high concentration of carbohydrates, which can facilitate the multiplication of yeasts, molds, and enteric bacteria. The presence these microorganisms cause losses in silage process of sorghum.

In general, the fermentation losses imply in the reduction of the availability of the ensiled forage, since there is no way to recover the DM losses in the form of gases and effluent.

The exposure of silage to air, converting the anaerobic environment (responsible for the conservation of forage) to aerobic, can cause changes in its chemical composition, altering its nutritional value, because the population of microorganisms that were dormant (bacteria, yeasts, and then mold action) and with oxygen began intense metabolic activity [35].

There is reduction in soluble components of silage, which are used as substrates for these microorganisms [30] and may even be a degraded part of the fibrous portion of food by fungal microbiota [28].

Evaluation of the aerobic stability of sorghum silages [26] found the aerobic deterioration losses of 85.6 kg/t DM in silages upon exposure to air during 48 hours. As the air to silage exposure is unavoidable during discharge, many research studies aim to reduce the aerobic deterioration with the use of additives [36].

The adding urea to acidic silage can neutralize part of acidity in the chemical reaction by partial neutralization, where, in an acid environment, an agent that has alkalizing action forms salts of organic acids [37] and subsequently providing the nitrogen applied [24].

Chemical additives such as urea can also benefit from the silage sorghum (**Table 4**). Although sorghum silage with urea present pH values and higher N ammonia, it does not mean that the

TRAT	pH	NH_3/TN	LA	AA	ET
Sorghum	4.73	0.228	6.01	0.83	0.44
Sorghum + LB	4.78	0.257	3.09	3.93	0.46
Sorghum + LP	4.25	0.189	12.46	0.56	0.40
Sorghum + 0.5% urea	3.69	0.169	5.27	0.85	0.60
Sorghum + 1.0% urea	3.73	0.401	6.69	0.37	0.26
Sorghum + 2.0% urea	3.76	0.525	5.71	0.70	ND
Sorghum + 4.0% urea	3.98	0.767	6.72	0.83	ND

Note: LB = *Lactobacillus buchneri;* LP = *Lactobacillus plantarum;* ND = Not detected.
Sources: Adapted from Filya [38] and Santos [26].

Table 4. Values of pH, relation ammoniacal nitrogen/total nitrogen (NH_3/TN), lactic acid (LA), acetic acid (AA), and ethanol (ET) of sorghum silage.

fermentation process is undesirable. Urea may act primarily in the metabolism of microorganisms, such as yeasts, reducing the conversion of the soluble compounds to ethanol, reducing DM losses. Furthermore, the addition of urea in sorghum silage had no negative effect on the production of lactic acid [26].

The sorghum has high WSC that can excessively acidify the silage due to excessive lactic acid production. The effect of different doses of urea on sorghum silage [26] found that the addition of urea reduced DM and WSC losses, reducing the production of ethanol from treated silages. Another benefit noted by the author was a high possibility of recovery of the nitrogen applied in the silages by incorporating the biomass ensiled.

The use of microbiological and chemical additives in sorghum silage can benefit from the fermentation process, and prolong the aerobic stability of silages [26].

After evaluating sorghum silage inoculated with lactic acid bacteria homofermentative and heterofermentative (**Table 4**), the researchers observed that the pH and WSC concentration decreased during fermentation, while increased lactic acid, acetic acid, ethanol, and ammonia content [38].

The addition of inoculants from lactic acid bacteria, such as *Lactobacillus buchneri* and *Lactobacillus plantarum* can benefit fermentation. Sorghum silages additive with *L. plantarum* showed low pH, lower content of acetic acid, ammonia nitrogen, and increased the production of lactic acid [38]. While silage inoculated with *L. buchneri* had a higher content of acetic acid and ethanol and lower lactic acid concentration [38].

L. buchneri is heterofermentative bacteria capable of converting water-soluble carbohydrates into lactic acid and other compounds with less acidifying power of the medium, such as acetic acid [39]. Still, these bacteria are capable of producing ethanol, which justifies higher values in the silage [39].

Another alternative is production of sorghum silage mixed with grasses. The sorghum silage has a high carbohydrate concentration, which implies the production of acid silage with predisposition to the development of deleterious microorganisms such as yeasts, and when under aerobic conditions in the silo-opening phase, aerobic stability is reduced.

In turn, grasses silages have lower amounts of WSC, buffering capacity, and relatively larger pH, which would lead to an increase in the production of acetic acid, the resulting product is essentially heterofermentative bacteria. Acetic acid has antifungal properties and may delay the development of fungi and degradation of nutrients in silage with high nutritional value, thus increasing the aerobic stability.

Considering these characteristics, the production of mixed silage sorghum with grass could promote appropriate fermentation profile, resulting in silage quality, as well as increase the aerobic stability of silage when exposed to air in the discharge phase, resulting in the reduction of aerobic degradation losses.

Evaluating sorghum silage mixed with 0, 25, 50, 75, and 100% of elephant grass, researchers found losses are reduced by gases (up to the level of 50%) and effluent (when added 75% grass elephant) in sorghum silage mixed with elephant grass [40]. Still mixed with elephant grass silages showed high resistance to heating after exposure to air of silage, there was an improvement in the aerobic stability of silage.

4.2. Pearl millet

The pearl millet (*Pennisetum glaucum*) is a grass of tropical region that can be considered alternative to forage production in Brazilian semiarid because it is a short cycled plant with high nutritive value adapted to climatic and soil conditions and it has great potential of production [41]. Because of its hardiness, rapid growth, adaptation to low soil fertility, and excellent biomass production capacity, it is an alternative to semiarid climates, where there are large climatic uncertainties.

This grass species has been widely used by producers as an alternative to attempt the requirements of animals in the critical part of the year. Pearl millet has been used as forage for the production of silage in periods of drought because of its specific characteristic such as more persistent drought, adapted to low fertility soils, fast growth, and good biomass production [35, 42].

Researchers evaluated the recovery of dry matter and losses of dry matter in the form of gases and effluent, and pH in silage of two pearl millet genotypes under nitrogen fertilization and found that the silages with lower pH were decreased the DM recovery and increased the soluble carbohydrates, which triggered the alcoholic fermentation [16].

The release of effluent can contribute to significant losses in the silage, considering that the DM content of pearl millet plants is relatively low. In many cases, good results have been achieved by using moisture-absorbing additives.

The incorporation of substances that absorb moisture inside the silo, such as citrus pulp, corn disintegrated with straw, corn cornmeal, and sorghum, favors the fermentation process. The incorporation of 3–7% of additives is sufficient to increase the DM content of the silage up to 25% DM, but this strategy should always be evaluated based on cost. Another alternative is to prewilting of the forage to be ensiled. This practice is effective. However, due to the significant increase in hand-to-work has proved more viable for small-scale silage production.

4.3. Grasses

Grasses cultivated under tropical conditions have high production in favorable season and reduction in unfavorable periods. Usually, there has been a fodder surplus in times of water, which should be maintained for subsequent supply in the drought period of the year. In this context, the grasses surplus ensiling can be a good practice to increase the supply of dry matter to animals in unfavorable times. Nevertheless, grasses have low DM and WSC content, as well as a limited number of indigenous bacteria, so that they require the techniques that increase their DM content and favoring the production of lactic acid bacteria [43].

The tropical grasses have low dry matter content, high power buffer, and low in soluble carbohydrates in the growth stages that have adequate nutritional value, which may harm the conservation process through the silage due to the possibilities of arising secondary fermentations, increasing the losses, and reducing the final quality of the ensiled material [44].

Researchers evaluated the effect of plant maturity on the DM content [45] and found the DM contents of 19.42, 21.06, 20.25, and 22.41% for 30 crops with heights of 40, 50, and 60 cm, which are unfavorable for appropriate fermentation of grass silage.

The WSC content in grasses is generally low depending on species and time of harvesting. The minimum WSC concentration to ensure the appropriate fermentation process is in the range 8–10% (DM basis) [13]. The WSC represents the main substrate for lactic acid bacteria, and must be at high concentration in plants prior to ensiling, so that the fermentation process is accelerate and the pH lowered rapidly, thereby inhibiting the growth of undesirable microorganisms.

The WSC and DM contents and buffering capacity influence directly the fermentation process of silage. Researchers [46] found that the DM and WSC content increases with the increase of regrowth age. Water-soluble carbohydrate levels in tropical grasses are low and thus it is difficult to reduce pH because of the absence of substrate for lactic acid bacteria, which suppresses the fermentation process.

Besides WSC and DM contents, buffering capacity also influences the ensiling process. The buffering capacity of forage resists changes in pH, which reduced the rapid lowering of pH necessary for forage preservation. The ratio of WSC and buffering capacity is important for the silage process. When the ratio is decreased it needs to increase in the DM content to avoid undesirable fermentation inside the silo.

The control of the ensiling process may be realized by the use of additives. Researchers [47] evaluated the effect of citrus pulp on Tanzania grass silage and found increased ratio of WSC and buffering capacity, which resulted in improved fermentation characteristics of silages with reduction of pH and ammonia-N values.

Another way to increase the level of soluble carbohydrates of forages before ensiling is the inclusion of sugarcane. The benefits of using sugarcane are similar to molasses to increase the

WSC content, resulting in reduction of pH and ammonia-N concentration and increasing the DM content [48].

Other sources that are used as additives, which are rich in soluble carbohydrates, are the residuals of fruit processing, such as cherry, pineapple, guava, passion fruit, mango, and papaya. These residues are usually dry, and used as both WSC sources and to increase the DM content of grass silage.

4.4. Leguminous

The leguminous species found in semiarid regions are drought tolerance. In order to reduce production costs, leguminous are often used as protein banks to feed ruminant animals, since the protein is expensive nutrient for animal nutrition [49].

The main leguminous fed to cattle in the semiarid region are leucaena (*Leucaena leucocephala*), pigeon pea (*Cajanus cajan*), Gliricidia (*Gliricidia sepium*), jitirana (*Merremia aegyptia*), sisal (*Agave sisalana*), perennial peanut (*Arachis pintoi*), among others.

Although these species are widely used as protein bank, some species of leguminous produced in the semiarid region have antinutritional compounds such as cyanide and tannin. These compounds may have a negative effect on ruminal degradation and become toxic when leguminous are present in excess. The ensiling process can soften or remove these undesirable compounds, improving the quality of food that provides to animals. This process has often been used for feeding animals in feedlot [50].

Leguminous species are not favorable for silage because of low concentrations of dry matter and water-soluble carbohydrates, and high protein content and buffering substances [51]. Because the amount of soluble carbohydrates, DM content, and buffer capacity [39], the fermentation process of leguminous silage may not be acceptable. However, the use of additives can improve the silage fermentation of these leguminous.

The fermentation of the silage leguminous is resistant to pH reduction due to the high buffering capacity and low content of soluble carbohydrates, which makes the highest production of lactic acid. There are a high number of pulses present in semiarid. Thus, it is important to use techniques which aimed at improving the ensiling process of legumes, making it favorable for silage.

The dry matter content directly influences the fermentative activity [13]. High moisture content and buffering capacity associated with low soluble carbohydrate content can lead to increased butyric fermentation, with losses of nutrients in the final food.

Leguminous have a high content of protein and minerals. Salts of organic acids, sulfate, nitrates, chlorides, and orthophosphate form the anion fraction of forage, which correspond approximately 68–80% of buffer capacity [52]. The disadvantages of leguminous silage are the need for increased lactic acid production to compensate for the high buffering capacity and reduce the pH to values below 4.0 [53].

Some strategies are used which can modify and improve the fermentation process of leguminous ensiling. In **Table 5**, we found that the silage pH perennial peanut had reduced after the

TRAT	pH	NH_3/TN	LA	AA	BA	PA
Perennial peanut (PP)	5.48a	18.22a	0.67h	0.09c	1.21a	1.61a
PP + 5% corn meal	4.76c	11.70ab	0.64h	0.17c	0.65b	0.86b
PP + 10% corn meal	4.57c	8.06bcd	2.29e	1.74ab	0.20de	0.86b
PP Wilted (PPW)	4.70c	4.15cd	1.10f	0.60bc	0.04e	0.03b
PP + Inoculant	5.18b	14.04ab	0.21i	3.25a	0.34cd	0.39b
PPW + Inoculant	4.67c	3.93cd	0.86g	1.15bc	0.03e	0.02b

Note: Means followed by the same letter in the column do not differ by 5% Tukey test.
Source: Adapted from Paulino et al. (2009).

Table 5. Values of pH, relation ammoniacal nitrogen/total nitrogen (NH_3/TN), lactic acid (LA), acetic acid (AA), butiric acid (BA), and propionic acid (PA) of perennial peanut silage.

addition of corn meal. Furthermore, additive increased the amount of lactic acid and acetic acid and reduced the content of ammonia nitrogen, butyric acid, and propionic acid. The additive corn meal positively changed the fermentation process of silage perennial peanut.

Other techniques such as wilting reduce losses in silage legumes. The wilting reduces the formation of organic ions that can result in the buffering effect on the silage fermentation process [54]. In **Table 5**, we confirmed the effect of wilting on silage perennial peanuts. Wilting reduced the pH, ammonia nitrogen content, butyric acid, and propionic acid, and increased the amounts of lactic and acetic acid. These changes are desirable, since lactic acid has preservative effect on the fermentation of silage to acidify [13].

The biological additives can be used in leguminous silage. **Table 5** shows the results of the addition of inoculant in perennial peanuts silage, when the wilting before ensiling occurred. This can be explained by the fact that due to the lower moisture content in the forage activity of lactic acid bacteria is increases and reduced the activity of other bacteria, such as clostridia, which are sensitive to osmotic pressure.

4.5. Cactus pear

The cactus pear (*Opuntia ficus-indica* and *Nopalea cochenillifera* Salm Dyck) has been increasing in the face of constant climate changes in the current production scenario [55] and its use in the objective Brazilian semiarid minimize the action of seasonality in the production process, providing energy and increasing the availability of water via food for animals.

In order to rationalize the use of this forage resource, the cactus pear as a silage is an alternative to this region. From the productive point, and the conservation of the nutritional value of the forage, the cactus pear silage maximizes the use of natural resources found in the Brazilian semiarid, enabling ranchers a new alternative for conservation of foods rich in water and energy, which adds more value to this Cactaceae in arid and semiarid regions.

Cactus pear has a low DM content and high WSC content, which could lead to alcoholic fermentation. However, researchers [56] evaluated cactus pear silages added with urea and found appropriate fermentation and low nutrient losses in silage. Despite some unfavorable attributes for silage, other characteristic of the cactus pear as per their bioactive compounds must be taken into consideration.

During rainy seasons, the cactus pear crop is not recommend for the ensiling process, because of the high moisture content that may bring difficulties in handling this material.

Other aspects related to fermentation kinetics of cactus pear silage are the percentage of organic acids found in the cactus pear cladodes, such as oxalic, citric, malonic succinic, and tartaric acids [57], which buffers the environment that impedes the lowering of pH.

Cactus pear is forage with low DM content and high WSC concentration, which may favor the development of undesirable fermentation. However, the bioactive compounds present in cactus pear promote homeostatic conditions in ensiled mass.

The emulsifier gel is formed after cutting of cactus pear, resulting of breaking of chlorenchyma and parenchyma cells, it is store mucilage, a hydrocolloid that promotes fluid retention. The hydrocolloids are compounds formed by highly hydrophilic polysaccharides, which reduce the movement of water providing increased viscosity of materials and thus the mucilage formation [58]. These compounds may be responsible for reducing effluent losses due to mucilage aggregates of fluid compounds.

The interaction of forage characteristics and its associative effects, as well as the handling, during ensiling directly influence the efficiency of the preservation process. The additives, in general, have been test more often in order to facilitate the practice of forage silage with high moisture and WSC content. The reports evaluating the silage cactus pear are still incomplete, as well as studies indicating additives for silage.

In recent studies with silage palm, researchers [56] conducted experiments to evaluate the losses resulting from the fermentation of forage cactus pear using additives such as urea and wheat bran. It observed that the urea reduced the effect of the increasing DM content and the crude protein values of cactus pear silage.

The cactus pear has favorable characteristics for the ensiling process; it is possible to produce good quality silage. Although many believe that the characteristics of the cactus pear, especially high WSC content, imply in inadequate fermentation characteristics. Cactus pear consists of elements that make it potential to be used as silage. Still, cactus pear silage is composed of a diet rich in energy for ruminants, as well as serve as an alternative source of metabolic water readily available in animal feed, especially in times of drought.

5. Final considerations

The use of plant to appropriate silage in combination with cultivate, harvesting, and silo filling results in a successful preservation of forage as silage.

Tropical crops, due to the tolerance of low water availability, are ideal for preserving forage as silage. In semiarid regions, the fermentative process of forages varies with conditions, and sometimes it requires additives.

Author details

João Paulo Farias Ramos[1]*, Edson Mauo Santos[3], Ana Paula Maia dos Santos[3], Wandrick Hauss de Souza[1,2] and Juliana Silva Oliveira[3]

*Address all correspondence to: jpemepapb@yahoo.com.br

1 Empresa Estadual de Pesquisa Agropecuária, Paraíba, Brazil

2 State Company of Agricultural Research of Paraíba, João Pessoa PB, Brazil

3 Federal University of Paraiba, Areia PB, Brazil

References

[1] Sampaio, E. V. S. B. Characteristics and potential. In: Gariglio, M. A. et al. (eds.). Sustentable use and conservation of forest resources of caatinga. 2nd ed. Brasília: Serviço Florestal Brasileiro, 2010. pp. 29–42.

[2] Menezes, R. S. C. et al. Cactus pear productivity in rural properties. In: Rômulo, S. C. et al. (eds.). Palm in Northeast Brazil: current knowledge and new perspectives of use. Recife: Ed. Universitária da UFPE, 2005. 258 p.

[3] Lima, C. D. S., Gomes, H. S., Detoni, C. E. Adding urea and *Saccharomyces cerevisiae* yeast in the protein enrichment of forage cactus pear (*Opuntiaficus indica* L.) cv. miúda. Magistra. V. 16, n. 1, pp. 01–08, 2004.

[4] Mitra, J. Genetics and genetic improvement of drought resistance in crop plants. Current Science. V. 80, n. 6, 2001.

[5] Blum, A. Drought resistance, water-use efficiency, and yield potential – are they compatible dissonant, or mutually exclusive? Australian Journal of Agricultural Research. V. 56, n. 11, pp. 1159–1168, 2005.

[6] Levitt, J. Responses of plants to environmental stresses. New York: Academic Press, 1972.

[7] Lopes, M. S. et al. Enhacing drought tolerance in C4 crops. Journal of Experimental Botany. V. 62, n. 9, pp. 3135–3153, 2011. Doi:10.1093/jxb/err105

[8] Yancey, P.H. Living with water stress: evolution of osmolyte systems.Science,V. 217, n. 4566, pp.1214–1222, 1982.

[9] Gaff, D.F. Adaptation of plants to water and high temperature stress. Turner, N.C and Kramer, P.J (eds). New York: Wiley, pp. 207–230, 1980.

[10] Edwards, G. E., Franceschi, V. R., Voznesenkaya, E. V. Single cell C4 photosynthesis versus the dual-cell (kranz) paradigm. Annual Review of Plant Biology. V. 55, pp. 173–196, 2004.

[11] Robertson, M. J. et al. Physiology and productivity of sugarcane with early and midseason water deficit. Field Crops Research. V. 64, pp. 211–227, 1999.

[12] Xoconostle-Cazares, B. et al. Drought tolerance in crop plants. American Journal of Plant Physiology. V. 5, n. 5, pp. 241–256, 2010.

[13] McDonald, P., Henderson, A. R., Heron, S. The biochemistry of silage. 2nd ed. Marlow: Chalcombe. 1991. 340 p.

[14] Ferreira, J. J. Corn maturation stage and sorghum ideal for silage. In. Cruz, J. C. et al. (Eds.) Production and use of corn silage and sorghum. Sete Lagoas: Embrapa Milho e Sorgo, 2001, pp. 405–428.

[15] Driehuis, F., Van Wikselaar, P. G. V. The occurrence and prevention of ethanol fermentation in high dry matter grass silage. Journal of the Science of Food and Agriculture. V. 80, pp. 711–718, 2000.

[16] Pinho, R. M. A. et al. Silages of pearl millet submitted to nitrogen fertilization. Ciência Rural. V. 44, n. 5, pp. 918–924, 2014.

[17] Weinberg, Z. G., Chen, Y. Effects of storage period on the composition of whole crop wheat and corn silages. Animal and Feed Science Technology. V. 185, pp. 196–200, 2013.

[18] Pereira, O. G., Rocha, K. D., Ferreira, C. L. L. F. Chemical composition, characterization and quantification of the population of microorganisms in elephant grass cv. Cameroon (*Pennisetum purpureum*, Schum.) and silages. Brazilian Journal of Animal Science. V. 36, n. 6, pp. 1742–1750, 2007.

[19] Kung Jr., L. A review on silage additives and enzymes. 2002. Disponível em: www.ag.udel.edu/department/anfs/faculty/kun…/a_review_on_silage_additivies_and.html [Accessed on 25 June 2014].

[20] Pahlow, G. et al. Microbiology of ensiling. In: Silage science and technology. Madison. Proceedings…Madison: ASCSSA-SSSA, Agronomy. V. 42, 2003. pp. 31–93.

[21] Santos, E. M., Zanine, A. M. Tropical grasses silage. Colloquium Agrariae. V. 2, n. 1, pp. 32–45, 2006.

[22] Van Soest, P. J. Nutritional ecology of the ruminant. Ithaca: Cornell University Press. 2nd ed. 1994. 476 p.

[23] Itavo, L.C.V. et al. Chemical composition and fermentative parameters of elephant grass and sugar cane with additive. Brazilian Journal Animal Health and Production, v. 11, n. 3, pp. 606–617, 2010.

[24] Schmidt, P. et al. Chemical and biological additives in sugar cane silage. 1. Chemical composition of silages, intake, digestibility and feeding behavior. Brazilian Journal Animal Science. V. 36, n. 5, pp. 1666–1675, 2007 (supplement).

[25] Santos, E. M. et al. Chemical composition, losses and fermentation profile of elephant grass silages with inclusion levels of jackfruit. Journal of Animal Health and Production. V. 9, n. 1, pp. 64–73, 2008.

[26] Santos, A. P. M. Sorghum silage BRS Ponta Negra with urea. 2013, 57 f. Dissertation (Masters in Animal Science) – Federal University of Paraiba, Areia.

[27] Neumann, M. et al. Characteristics of the fermentation of silage obtained in different silos under the effect of particle size and time of harvest of corn plants. Rural Science. V. 37, n. 3, pp. 847–854, 2007.

[28] Guim, A. et al. Aerobic stability of silage of elephant grass (*Pennisetum purpureum*, Schum) wilted and treated with microbial inoculant. Brazilian Journal Animal Science. V. 31, n. 6, pp. 2176–2185, 2002.

[29] Danner, H. et al. Acetic acid increases stability of silage under aerobic conditions. Applied and Environmental Microbiology. V. 69, n. 1, p. 562–567, 2003.

[30] Vieira, F. A. P. et al. Quality sorghum silages with additives. Brazilian Archives of Veterinary Medicine and Animal Science. V. 56, n. 6, p. 764–772, 2004.

[31] Botelho, P. R. F. et al. Sorghum genotypes Evaluation first cutting and regrowth for the production of silage. Brazilian Journal of Maize and Sorghum. V. 9, n. 3, pp. 287–297, 2010.

[32] Pinho, R. M. A. et al. Use efficiency rain for sorghum silage genotypes on Paraiba Semi-Arid. In: Congress animal production Northeastern, Vol. 6, 2010, Mossoró, Anais Mossoró: CNPA, 2010.

[33] Neumann, M. et al. Evaluation of different sorghum hybrids (*Sorghum bicolor*, L. Moench) as the components of the plant and produced silages. Brazilian Journal of Animal Science. V. 31, n. 1, pp. 302–312, 2002.

[34] Fernandes, F. E. P. et al. Sorghum silage with added urea in two periods of storage. Brazilian Journal of Animal Science. V. 38, n. 11, pp. 2111–2115, 2009.

[35] Amaral, P. N. C. et al. Nutritive quality of silage of three varieties of millet. Science and Agrotecnology. V. 32, n. 2, pp. 611–617, 2008.

[36] Jobim, C. C. et al. Methodological advances in evaluation of preserved forage quality. Brazilian Journal of Animal Science. V. 36, p. 101–119, 2007 (supl. special).

[37] Lopez, J. et al. Effect of nitrogen source, stage of maturity, and fermentation time on pH and organic acid production in corn silage. Journal of Diary Science. V. 53, pp. 1225–1232, 1970.

[38] Filya, I. The effect of *Lactobacillus buchneri*, with or without homofermentative lactic acid bacteria, on the fermentation, aerobic stability and ruminal degradability of wheat, sorghum and maize silages. Journal of Applied Microbiology. V. 95, pp. 1080–1086, 2003.

[39] Oude Elferink, S. J. W. H. et al. Anaerobic conversion of lactic acid to acetic acid and 1,2-propanediol by lactobacillus buchneri. Applied and Environmental Microbiology. V. 67, n. 1, pp. 125–132, 2000. DOI: 10.1128/AEM.67.1.125–132.2001

[40] Ramos, R. C. S. Silages mixed assessment of elephant grass with sorghum.2014. Dissertation (Masters in Animal Science) – Federal University of Paraiba, Areia.

[41] Kollet, J. L., Diogo, J. M. S., Leite, G. G. Forage yield and chemical composition of varieties of pearl millet (*Pennisetum glaucum* (l). R. Br.). Brazilian Journal of Animal Science. V. 35, n. 4, pP. 1308–1315, 2006.

[42] GuimarãeS J. R. R. et al. Dry matter, crude protein, ammonia nitrogen and pH of silages three pearl millet genotypes [Pennisetum glaucum (l). R. Br.] At different periods of fermentation. Brazilian Journal of Maize and Sorghum. V. 4, n. 2, pp. 251–258, 2005.

[43] Santos, D. C. et al. Nitrogen and phosphorus levels in forage cactus (Opuntia ficus-indica) IPA-20 clone in two spacings. In: Congress animal production Northeast, Vol. 4, 2006, Petrolina, Anais Petrolina: SNPA, 2006. pp. 381–383.

[44] Evangelista, A. R. et al. Production of silage marandu grass (*Brachiaria brizantha* Stapf cv. Marandu). Agrotecnica Science. V. 28, n. 2, pp. 443–449, 2004.

[45] Pinho, R. M. A. et al. Microbial and fermentation profiles, losses and chemical composition of silages of buffel grass harvested at different cutting heights. Revista Brasileira de Zootecnia. V. 42, n. 12, pp. 850–856, 2013.

[46] Ribeiro Junior, G. O., Agronomic characteristics and quality of grass silages *Andropogon gayanus* in four cutting ages and fermentation profile of silages after 56 days of growth. 2009, 46 f. Dissertation (Masters in Animal Science) – School of Veterinary Medicine, Federal University of Minas Gerais, Belo Horizonte.

[47] Ávila, C. L. S. et al. The levels of soluble carbohydrates of Tanzania grass ensiled with additives assessment. Brazilian Journal of Animal Science. V. 35, n. 3, pp. 648–654, 2006.

[48] Santos, G. et al. Pasture characterization of buffel grass deferred and cattle diet during the dry period in Pernambuco. Brazilian Journal of Animal Science. V. 34, n. 2, p. 454–463, 2005.

[49] Chen, C. P. et al. Fodder trees and fodder shrubs in range and farming system of the Asian and Pacific region. In: Legume trees and other fodder trees as protein sources for livestock.1991.

[50] Costa, C. X. Nutrients intake, production and performance characteristics of carcass santa ines lamb containment in aloft sergipe semiarid. 2008, 68 p. Dissertation (Masters in Animal Science) – Federal University of Paraiba, Areia.

[51] Leonel, F. P. et al., Consortium signal grass and corn: crop yields and quality characteristics of silages made with plants of different ages. Brazilian Journal of Animal Science. V. 37, n. 11, pp. 2031–2040, 2008.

[52] Playne, M. J., McDonald, P. The buffering constituents of herbage and of silage. Journal of the Science of Food and Agricultural. V. 17, pp. 262–268, 1966.

[53] Lavezzo, W. Elephant grass ensilage. In: Symposium pasture management, Vol. 10. Piracicaba: Foundation of Agrarian Studies Luiz de Queiroz, pp. 169–275, 1993.

[54] Ribeiro, J. L. Tanzania and Marandu grass silages evaluate for storage losses, fermentation profile nutritional value, performance of animals in the presence of chemical additives, microbial and absorbing moisture sources. Thesis (Agronomy doctoral degree). School of Agriculture, Luiz Queiroz University of São Paulo, Piraciba, 2007.

[55] Silva, A. P. G. Regrowth assessment maniçoba depending on planting density, organic and mineral fertilizers. 2010, 60 f. Dissertation (Masters in Animal Science) – Centre of Agricultural Sciences/Federal University of Paraiba, Areia.

[56] Nogueira, M. S. Fermentative profile and chemical composition of cactus forage silages added with urea and wheat bran. 2015, 60 f. Dissertation (Masters in Animal Science) – Federal University of Paraiba, Areia.

[57] Stintzing, F. C., Carle, R. Cactus stems (*Opuntia* spp.): A review on their chemistry, technology and uses. Molecular Food and Nutrition Research. V. 49, pp. 15–194, 2005.

[58] Saenz, C. et al. Opuntia spp mucilage's: a functional component with industrial perspectives. Journal of Arid Environments. V. 57, pp. 275–290, 2004.

Chapter 5

Potential Use of Nonconventional Silages in Ruminant Feeding for Tropical and Subtropical Areas

Jaime Salinas Chavira

Additional information is available at the end of the chapter

http://dx.doi.org/10.5772/64382

Abstract

The conventional silage uses crops such as corn, sorghum or other forages for this specific objective. The nonconventional silages use by-products, co-products and other materials obtained during the harvest or during the processing in the industry of sugarcane, juice extraction of citrus, pineapple, cassava, pumpkin and others. These products are available in high amounts during a short period of time. These by-products can be ensiled to maintain their nutritive value during longer period in the year and then used as feed for animals. These by-products have adequate characteristics for ensiling, i.e., moisture content and fermentable carbohydrates. Forages reduce their crude protein (CP) concentration in a period of the year (dry season or in winter), which may limit animal production. Most by-products used for silage have low CP concentration; some additives may help increase the nutritive value of these silages. These by-products (sugarcane, juice extraction of citrus, pineapple, cassava, pumpkin and others) can be mixed and ensiled with other by-products as poultry excreta or forage rich in protein to obtain silage with greater CP concentration. The research shows the feasibility of obtaining good quality silages from sugarcane tops, by-products of citrus, cassava and pumpkin; the particularities of each are discussed in detail in this chapter.

Keywords: potential use, nonconventional, by-products, silage, tropics

1. Introduction

The preparation of conventional silage considers crops for this specific purpose; these crops include corn, sorghum or other forages. Nonconventional silages use by-products, co-products and other materials different from conventional crops; they include by-products of sugarcane, juice extraction of citrus, pineapple, cassava, pumpkin and others. These products are available in high amounts during a short period of time, the harvest season. Their preservation is required so they can be maintained for longer periods of time and used then as feed for animals. Silage represents an appropriate technique for this purpose. Also these materials may have adequate characteristics for ensiling, i.e., adequate moisture content and high fermentable carbohydrates. These byproducts may have low crude protein (CP) and mineral concentration and some additives may help increase the nutritive value of these silages.

The improvement in nutritional quality of silages with nonconventional products may contribute to better animal feeding and production in tropical and subtropical areas. In addition, the use of alternative silages for animal feeding also may contribute by reducing environmental pollution. If by-products are not used in a short time (during the harvest season) they will be wasted in fields and in other cases they are burned. The objective of this chapter is to review the potential use of nonconventional silages (sugarcane tops [SCT], citrus byproducts, pumpkin and cassava) for ruminant feeding in tropical and subtropical areas.

2. Ensiling process

Silage is the preservation of feeds by anaerobic fermentation, usually by epiphytic bacteria that convert soluble carbohydrates mainly to lactic acid, and minor amounts of other volatile fatty acids. This reduces pH, which inactivates or inhibits microbial growth and results in the preservation of ensiled material. The ensiling process has four stages. In phase 1, aerobic microorganisms are active during the aerobic phase and occur under aerobic conditions during the few hours after ensiling. The ensiled material and facultative microorganisms (yeasts and enterobacteria) continue respiration, reducing the oxygen present. The enzymes of the ensiled material are active and pH is close to 6. In phase 2, anaerobic fermentation starts and continues for several days depending on substrate availability and ensiling conditions. Lactic bacteria become the main strain, and lactic acid reduces pH to 3.8–5.0. In phase 3, the process is stable, because changes can occur even in anaerobic conditions; most microorganisms reduce their numbers. During this phase, while the silage maintains anaerobic conditions, the process is practically unchanged. Phase 4 starts with the opening of the silage, or air exposure. The spoilage of silage in this phase is due to two processes: one is the degradation of acids that preserve the silage and the second is the spoilage by some microorganisms [1].

To produce good quality silage the following principles of fermentation during the phases of ensiling should be considered. For phase 1, adequate particle size of ensiled material with efficient filling (adequate packing density) will reduce aerobic respiration that allows faster growth of anaerobic microorganisms that produce lactic acid. Phases 2 and 3 can be enhanced and/or stabilized using some additives to silage during its preparation. For fast time of pH reduction and decrease dry matter (DM) losses, additives containing water soluble carbohydrates are used. To improve the fermentation process, some microbial inoculants, organic acids and enzymes can also be used. It is pointed out [2] that additives in silage stimulate lactic acid bacteria growth, responsible of silage stability, decreasing nutrient loss during fermentation and resulting in silage of higher nutrient concentration. Additives that contribute to silage stability are acetic, propionic and caproic acids; also ammonia and some inoculants may contribute to silage stability [2]. To reduce the spoilage of ensiled material in phase 4, it is recommended that the silage be used as fast as possible once the silo is opened [1].

3. Nutritive value of silages

3.1. Nutritive value of sugarcane tops silage

In tropical areas, high amounts of vegetal biomass are produced due to the dynamic ecosystem, which is favored by the climatic conditions, i.e., humidity and temperature that propitiate accelerated growth of plants. An example of these plants is sugarcane (*Saccharum officinarum*); it can be fed to animals as an entire plant [3]. During the harvest of sugarcane for sugar extraction in the field an abundant biomass of sugarcane tops is wasted or burned; they constitute about 15% dry matter of total plant [4] and have greater protein content than the stalks [5], representing an alternative forage for ruminants in subtropical areas, where the climatic conditions complicate forage preservation; for this, sugarcane tops silage represents an alternative. The replacement of stalks by fresh tops of sugarcane in feedlot cattle diets has increased dry matter intake and body weight gain, **Table 1** [6]. The supplementation with urea of slow ruminal degradation to fresh sugarcane tops improved weight gain in lambs, **Table 2** [7]. Ruminal fermentation of fresh sugarcane tops is improved when supplemented with nitrogen and nonstructural carbohydrates [8], also similar results are observed in **Table 3** [9].

	Top:stalk fresh basis					
	0:100	20:80	40:60	60:40	80:20	100:00
ADG, kg/d	0.605	0.614	0.699	0.760	0.788	0.839
DMI, kg/d	4.52	4.66	6.49	6.40	6.76	7.50
Feed/gain	7.47	7.59	9.28	8.35	8.57	8.94

ADG = average daily gain; DMI = dry matter intake; feed conversion = feed intake/weight gain.
Adapted with permission from Ferreiro and Preston [6].

Table 1. Summary of growth performance of feedlot cattle fed with different proportions of tops:stalks of sugarcane.

	SCT	SCT plus slow degrading urea	SCT plus slow degrading urea plus corn plant
ADG, g/d	70	135	218
DMI, g/d	474	797	917

SCT = sugarcane tops; ADG = average daily gain; DMI = dry matter intake.
Adapted with permission from Galina et al. [7].

Table 2. Summary of growth performance of feedlot lambs fed with sugarcane tops supplemented with slow degrading urea and corn plant.

	Ruminal parameters			
	Washing loss (A)	Degradability of water insoluble fractions (B)	Potential degradability (A + B)	Fractional degradation rate (C)
Trial 1				
0 g/kg TG	0.17	0.36b	0.54b	0.0490
300 g/kg TG	0.17	0.39a	0.57a	0.0496
Trial 2				
0 g/kg PM	0.17	0.38b	0.56b	0.0437
300 g/kg PM	0.17	0.40a	0.58a	0.0472
Trial 3				
0 g/kg Urea	0.17	0.38b	0.56b	0.0448b
24 g/kg Urea	0.17	0.42a	0.60a	0.0568a
Trial 4				
0 g/kg HNESO	0.17	0.37b	0.55b	0.0441b
1500 g/kg HNES15	0.17	0.45a	0.63a	0.0578a

TG = Taiwan grass (*Pennisetum purpureum*); PM = poultry manure; HNES = high nitrogen and energy supplement; the latter had (g/kg) ammonium sulfate 18, animal lard 40, cement kiln dust 16, corn 112, cottonseed meal 164, fish meal 42, limestone 32, mineral salts 10, molasses 182, orthophosphate 30, poultry manure 116, rice polishing 160, NaCl 40 and urea 38; within columns, different literals (a or b), denote statistical difference ($P<0.05$).
Adapted with permission from Ortiz-Rubio et al. [9].

Table 3. Parameters of ruminal kinetics of sugarcane tops supplemented with different feeds, data obtained from in situ incubations in steers.

Ensiling sugarcane tops is a logical alternative; however, this process may have complications. It is reported that ensiling reduced dry matter digestibility and feed intake in lambs [10], probably because of excessive production of ethanol during the process [11].

Values of 4.7% and 10.1% crude protein, 87% and 78% neutral detergent fiber (NDF), respectively, were reported for fresh and ensiled sugarcane tops [12]; however, this increase in protein could be a dilution effect and not by the fermentative process of ensiling. Acceptable color and odor, indicating no putrefaction was also reported; pH was from 4.0 to 4.04. In vitro gas production was higher for fresh than ensiled sugarcane tops at 24 h; however, the organic matter digestibility estimated from in vitro gas production was higher for ensiled sugarcane tops [12].

It was found that sugarcane tops (SCT) had lower CP and minerals than broiler litter (BL). These two feed ingredients can improve silage nutritional composition, fermentation characteristics, degradation of DM by microorganisms in the rumen and destruction of mycotoxin-producing fungi (MPF). Excessively high amount of BL can cause deleterious effects on the quality of the resulting silage product. It would therefore be recommended that a 30–45% inclusion rate is the most appropriate level of incorporation of BL in silages. Adequate levels of moisture are needed in silage [13]. From 30 to 45% BL enhanced lactic acid production and pH was acceptable; however, 60% of BL in silage resulted in high buffer capacity with high levels of ammonia production that caused silage pH increased.

In a study, sorghum stover was substituted with sugarcane top silage supplemented with urea [0 (T1), 5 (T2) and 10% (T3) DM] in high concentrate diets for feedlot hair lambs. It was observed a reduction of effective ruminal degradability with increased SCT contents in silage. Feedlot hair lambs observed reduced feed intake augmenting sugarcane tops silage in their ration. Nevertheless daily weight gain was not affected by diet. Feed efficiency (gain/feed intake) was not influenced by treatment. It was concluded that ensiled sugarcane tops constitute alternative forage in diets for growing-finishing feedlot lambs [14].

3.2. Nutritional value of citrus silage

Most citrus species are well adapted in tropical and subtropical areas. Citrus fruits are used as dessert, although considerable amounts are used for industrial juice extraction. Citrus production in the producing countries is increasing [15]. The augmented disposal costs in many parts of the world have stimulated attention in utilizing citrus by-product feedstuffs (BPFs) as alternate feeds for ruminants [16]. In ruminant feeding, the principal citrus by-products are fresh pulp, silage, dried, meal, molasses and citrus peel liquor. Other minor BPFs from citrus include cull or excess fruit. Citrus BPFs can be used as a high-energy feed in ruminant rations to support growth and lactation, with fewer negative effects on rumen fermentation than starch-rich feeds.

The world citrus production of the genus *Citrus* are sweet orange (*C. sinensis*: 67.8%), tangerine (*C. reticulata*: 17.9%), lemon (*C. limon*: 6.3%) and grapefruit (*C. paradisi*: 5.0%).

The remaining 3.0% of the *Citrus* genera are sour orange (*C. quarantium*), shaddock (*C. grandis*), citron (*C. medica*) and lime (*C. aurantifolia*). The largest world orange juice producing countries are Brazil, the United States, Mexico, Spain, China and Italy. Other significant orange producing countries include South Africa, Israel, Egypt, Iran, Cuba, Costa Rica, Belize, Japan and Australia [17]. It would be convenient to develop methods to preserve the fruit surplus during the production season in tropical countries that would enable this plant material to be utilized as animal feeds for longer periods of time [18].

It was showed that ensiling citrus by-products are possible; however, the high water content might affect the quality of the product [16]. This sense, citrus pulp silage produces high quality fermentation when straw and poultry litter are added [19]. In other research [20], fresh orange peel was ensiled without additive (control), or with enzyme inoculate (EI), formic acid (FA), propionic acid (PA) and acetic acid (AA). Samples of fresh and ensiled orange peel were analyzed for dry matter (DM), crude protein (CP), neutral detergent fiber (NDF), acid detergent fiber (ADF), *in vitro* dry matter and cell wall disappearance, pH, buffering capacity and ammonia N. In this study, it was concluded that the additives used did not improve nutritional value of orange peel silage. Similarly, it was observed that orange peel silage showed a high apparent digestibility, although additives did not alter nutritive value of silages [21]. In different results, three silages of orange fruit wastes (OFWs) combined with (1) urea (0.5%);

(2) leucaena (ratio 1:1 leucaena-OFW) and (3) orange fruit wastes plus a fish (*Oreochromis aureus*) of noncommercial size disintegrated (ratio 2:1). In this report, it was observed that orange fruit wastes plus leucaena silage showed high alcohol content. But the silage with fish had adequate chemical properties and could be recommended to farmers [22]. In the silage of kinnow mandarin (*Citrus nobilislour* × *Citrus deliciosa tenora*) fruit waste, it was observed that this by-product can be used to prepare good quality silage for goats [23]. In a similar report [24], the fermentative characteristics, intake, digestibility and aerobic stability of pineapple silage (PS) or citrus silage (CS) were studied. Crossbred rams were used to determine the *in vivo* digestibility. Final pH at 65 d was 3.21 and 3.32 for PS and CS; in both silages, population of Enterobacteriaceae was not detected. The DM and CP intakes and digestibility were similar among treatments. In this study, it was noted that both silages of by-products were unstable upon aerobic exposure, PS after 1 d when fermented 29 d and CS after 3 d when fermented 65 d. Results indicate that pineapple and citrus by-products could be preserved as silage and included in sheep diets at 20% substitution of grass without adverse results; however, they are susceptible to aerobic deterioration. This may represent the use of silage in feeding the animals as fast as possible when silage is open (**Table 4**).

	Days of fermentation			
	0	**7**	**29**	**65**
pH	5.4	3.7	3.5	3.3
Dry matter, %	24.2	19.7	19.0	18.7
Crude protein, %	5.7	5.9	6.4	6.5
NDF, %	16.2	23.4	24.2	23.7
Water soluble carbohydrates, %	5.6	1.8	4.2	5.4
Lactic acid, g/kg	0.02	1.17	1.2	1.7
Acetic acid, g/kg	0	0.20	0.27	0.36
Propionic acid, g/kg	0	0	0.01	0.01
Butyric acid, g/kg	ND	ND	ND	ND
Ammonia-N, g/kg	1.4	1.3	1.3	1.2

Adapted with permission from Pagán et al. [24]. NDF = neutral detergent fiber.

Table 4. Fermentation characteristics of citrus by-products silage at different periods of process.

3.3. Nutritive value of pumpkin silage

Fresh pumpkin can be fed to animals; they have seasonal availability. To preserve pumpkins and use them in different seasons of the year, silage may represent an alternative in ruminant feeding. The complete fruits are carbohydrate and protein rich. The total sugar and crude protein are 48.1% and 18.2%, respectively; however they have low dry matter concentration (16.8%). For this reason, when ensiling pumpkin, an adsorbent material should be included to have adequate water level. The silage of pumpkin with dried beet pulp has about 11% CP [25]. The nutritional quality of pumpkin silage was assessed by Halik et al. [26] who produced silages with chopped pumpkin (*Cucurbita maxima*) fruits mixed with dried beet pulp at an 80:20 ratio. Silages were prepared with or without inoculant; the inoculant contained *Lactobacillus plantarum* bacteria, endo-1,4-beta-glucanase, xylanase and glucoamylase. The inoculant was applied at 0.2% of the ensiled material. The material after ensiling (10 weeks) had lower crude fiber and ADF compared to fresh material, whereas the inoculant had no effect on silage pH (4.5 and 4.4 for no inoculant and with inoculant) but reduced ammonia, nitrogen and ethanol and increased lactic and acetic acids, indicating higher aerobic stability with improved silage quality.

Silages were made of pumpkin (*C. maxima*), sorghum straw, urea and cane molasses at levels of 73.4%, 25.6%, 1% and 0% (treatment 1); 72.13%, 21.87%, 0% and 6% (treatment 2); treatment 3 had 72.2%, 20.8%, 1% and 6%, respectively. With these three silages, three diets were formulated for lambs with similar concentrations of crude protein and metabolizable energy. With these silages, two trials were conducted. In trial 1, of growth performance, results showed similar weight gain, feed intake and feed efficiency of lambs fed the three diets. Based on these results, they concluded that silages of pumpkin with sorghum straw and additives can be used in ruminant feeding [27]. In trial 2, the apparent digestibility of complete diets for lambs containing the same silages was studied. The in vivo digestibility was of 81.27%, 70.7% and 70.31% for crude protein; 75.21%, 62.04% and 80.95% for ether extract (EE). These values were different ($P < 0.05$). The in vivo digestibility of dry matter and nitrogen-free extract was similar between the rations ($P > 0.05$). In this study, it was concluded that the digestibility was improved in ration with silage that contained only urea [28]. Although this study did not report fermentative characteristics of silages, urea might contribute for growth of bacteria that digested nitrogen.

Pumpkin can be cultivated for seed collection, with abundant residues that many times are wasted in field and have potential for ensiling and use in ruminant feeding. For this application, Hashemi and Razzaghzadeh [29] used pumpkin residue (PR; fleshy part of fruit that remains after seeds are collected). Pumpkin residue (71.4%) was mixed with wheat straw (28.6%) and ensiled with dry beet molasses (10% or 20%) and urea (0% or 5%). After 2 months, silages were evaluated for pH and dry matter. They concluded that PR may be ensiled with wheat straw as absorbent of moisture and beet molasses as fermentable additive.

In other research [30], the growth performance of male buffalo calves fed diets containing silage of pumpkin (*Cucurbita pepo*) residues was studied. Silages were prepared using pumpkin residues chopped at 2 cm; 700 kg of this was mixed with 300 kg of wheat straw and ensiled adding 100 l of a solution (10 kg urea plus 50 kg beet molasses in water). Pumpkin residual silage (PRS) replaced forage (alfalfa) at 0% (control), 20%, 40% and 60%. In this study, it is concluded that part of the alfalfa may be substituted with PRS at 60% level with no negative effects on male buffalo calves' fattening performance.

Pumpkin was ensiled (*C. maxima* D.) with dried sugar beet pulp (80:20 ratio) and studied its antioxidant potential. It was observed that ensiling increased the saturated fatty acid content and decreased the polyunsaturated content. Ensilage increased the polyphenol compounds and decreased the carotenoid and alpha-tocopherol content of the silages. Although there was a reduction of carotenoid and tocopherol compounds, the increase of polyphenol compounds suggests that the ensiling did not lower the silage antioxidant potential of pumpkins compared to fresh material [25].

3.4. Nutritive value of cassava silage

The importance of cassava (*Manihot esculenta* crantz) for livestock feeding was reported [31]. Cassava has high productivity per unit of land; it has low crude protein but a high amount of starch (about 85% DM basis). Different methods for ensiling this material have been explored. A study investigated the chemical composition and organoleptic traits of maralfalfa silage (*Pennisetum* sp.), containing 0%, 5%, 10% and 15% fresh cassava (*M. esculenta*). This research reported improved silage characteristics of maralfalfa grass with 15% of cassava, with acceptable pH; reduction of cell wall (NDF) fraction and the crude protein is maintained in the silage [32]. The fermentation characteristics of cassava silage at laboratory scale have been improved with microorganism-inoculant of genus *Lactobacillus*. This study reports reduction of ethanol and total VFA but maintained or increased lactic acid for adequate pH with inoculant treatment of silos; also it is reported that acid treatment goes to alcoholic fermentation. In addition, *Lactobacilli inoculum* generated homofermentative pattern [33]. In a similar study, *L. plantarum* and *Lactobacillus cellobiosus* increased the acidification rates of *M. esculenta* silages [34]. The results at laboratory silos are satisfactory; however, more conclusive results are required.

The chemical composition of cassava starch by-products before and after ensiling was studied; fermentation characteristics and growth of microorganisms were also determined. The results showed that ensiling reduced NDF and hemicellulose concentrations, but increased concentrations of ADF, cellulose and lignin. pH and microbial populations were reduced as the duration of silage fermentation increased. Predrying did not change the fermentative profile and microbiological population of silages at 28 and 56 d and reduced neutral detergent fiber and hemicellulose of silages. The wet waste residue silage showed a reduction in crude protein content in the course of the fermentation period. This research showed that cassava by-products have good fermentation characteristics [35]. Also, the fermentative characteristics and chemical composition of Elephant-Grass silages with cassava by-product (SM in relation to the grass fresh matter) was explored [36] . It was observed that in the level of 7.1% of SM addition, the silages had adequate dry matter content (30–35%) for a good fermentative process. In all levels of SM addition, the silages had appropriate pH values for silages (3.8–4.2). The cassava by-product up to the 20% level (on a grass fresh matter basis) at the elephant grass ensiling produced good fermentative characteristics and a better silage chemical composition. Although a minimum level of CP (7%) was not reached in any level of SM.

In another study [37], Holstein cows in diets were fed with silage of the residue from the extraction of cassava starch (SRECS), replacing 0%, 25%, 50%, 75% or 100% of the corn feed. Before ensiling the material had 128.0 g kg^{-1} of dry matter (DM), 25.3 g kg^{-1} of crude protein (CP), 25.0 g kg^{-1} of mineral matter (MM), 297.0 g kg^{-1} of neutral detergent fiber (NDF) and 6.1 g kg^{-1} of ether extract (EE) on a dry matter basis. After ensiling (40 d) the silage had

189.8 g kg^{-1} of DM, 24.4 g kg^{-1} of CP, 23.8 g kg^{-1} of MM, 324.9 g $kg^{-1}$of NDF, 271.9 g kg^{-1} of ADF and 05.4 g kg^{-1} of EE (dry matter basis). In this study, it was concluded that the silage of the residue from the extraction of cassava starch to replace the ground corn on feed negatively affects nutrient intake without changing the efficiency of milk production, milk composition or blood parameters of lactating cows.

Cassava by-product is starch-rich and promotes good fermentation characteristics; however, it is low in protein. The age was compared at harvesting time (7, 8 and 9 months) of cassava plants on whole crop silage quality. Plants were ensiled in laboratory silos. The results showed that ensiling reduced HCN content (more than 60%). Harvesting cassava plant at 8 months of age gave the best whole cassava plant silage quality (best physical characteristics and in vitro rumen digestibility). They also conclude that the low crude protein of cassava for ensiling could be improved mixing with other protein rich by-products like poultry litter [38]. In other research, silage of cassava (*M. esculenta*) by-product with poultry litter at 0%, 5%, 10%, 15% and 20% was produced. They observed that increasing levels of poultry litter influenced DM, CP, EE, ash, calcium, NDF, ADF, cellulose and hemicellulose content of the silage; however, no clear tendencies were found for lignin content. On the other hand, poultry litter addition decreased nonfibrous carbohydrate concentration and IVDMD of the silage. Although pH increased, the level was acceptable in all silages. The authors recommend 10% poultry litter to preserve nutritional and fermentative characteristics of the cassava silage [39].

Another alternative to improve crude protein in the cassava by-product is the inclusion of forage rich in protein. Cassava peels (CaPe) was ensiled with mixtures of *Gliricidia sepium* and *Leucaena leucocephala*; the nutritive value was assessed in goats. All diets were supplemented with molasses (40 g/kg) before ensiling which lasted 3 months. The silage with only CaPe (control) had the lowest hydrocyanic acid content. All silages had low pH (<4.5). Authors reported that *L. leucocephala* and *G. sepium* ensiled with CaPe did not affect fermentation but improved the CP content of the resulting silage. Increasing level of *L. leucocephala* reduced weight gain of animals. Silage of CaPe (control) improved weight of animals. They conclude that ensiling CaPe with foliages of *G. sepium* and *L. leucocephala* may be recommended, especially for the season of year when forages reduce availability and nutritive value.

During the harvest of root cassava, also can be collected the aerial part, the vegetative fraction containing mainly leaf may generate about 1.8 tons per ha of dry matter. Cassava leaf is protein rich; it contains about 21% CP [40]. The silage of cassava foliage with different levels of molasses was studied. Increasing molasses level did not influence DM, pH or lactic acid of silages; however, reduced CP and increased water soluble carbohydrates. Cyanic acid (HCN) was not influenced by molasses; however, all silages reduced HCN concentration after 2 months of fermentation. These results show the possibility of ensiling leaf cassava with low levels of molasses [41]. In other study, cassava leaves were used without additives, with molasses or with caged layer waste. All silages had adequate fermentation parameters. In this study, HCN was lower in silages with additives; the HCN (mg/kg) was of 112.3, 95.8, 84.7 and 89.3 for fresh leaves, silage of leaves, silages of leaves with molasses and silage of leaves with poultry excreta, respectively [42].

4. Conclusions

Silages of sugarcane tops, citrus, cassava and pumpkin represent an alternative in animal feeding; the particular characteristics of each should be considered for better silage production. The research shows the feasibility of producing good quality silages with these materials. This technology represents an alternative to enhance animal production, converting these products or by-products in good quality protein of animal origin. Research to improve the fermentation process during ensiling of these materials and their incorporation with other available resources must continue.

Author details

Jaime Salinas Chavira

Address all correspondence to: jsalinas@uat.edu.mx

College of Veterinary Medicine and Zootechnology, Autonomous University of Tamaulipas, Cd. Victoria, Tamaulipas, México

References

[1] Stefanie JWH, Elferink O, Driehuis F, Gottschal JC, Spoelstra SF. 1999. Paper 2.0: Silage fermentation processes and their manipulation. In: Silage Making in the Tropics with Particular Emphasis on Smallholders. Edited by 'tMannetje L. Available from: http://www.fao.org/docrep/005/x8486e/x8486e00.htm#Contents [accessed: 2016-02-01].

[2] Yitbarek MB, Tamir B. Silage additives: review. Open Journal of Applied Sciences 2014; 4: 258–274. http://dx.doi.org/10.4236/ojapps.2014.45026.

[3] Landell MGA, Campana MP, Rodrigues AA. The IAC 862480 variety as a new choice of sugar cane for forage purposes: production management and use in animal feeding. Technical Bulletin IAC193, Series Technological APTA 2002; 36 p.

[4] López I, Aranda EM, Ramos JA, Mendoza GD. Nutritional evaluation of eight sugar-cane varieties with forage potential. Cuban Journal of Agricultural Science 2003; 37: 375–380.

[5] Juárez LF, Vilaboa AJ, Díaz RP. 2009. Sugar cane (*Saccharum officinarum*): an alternative for substitution of corn (*Zea mays*) in the feeding of feedlot cattle. Available from: http://www.produccionbovina.com.ar/informaciontecnica/ invernada o engorde a corral o feedlot/69-cana azucar.pdf [accessed: 2016-02-01].

[6] Ferreiro HM, Preston TR. Fattening cattle with sugar cane: the effect of different proportions of stalk and tops. Tropical Animal Production 1976; 3: 131–138.

[7] Galina MA, Guerrero M, Puga CD. Fattening Pelibuey lambs with sugar cane tops and corn complemented with or without slow intake urea supplement. Small Ruminant Research 2007; 70: 101–109.

[8] Gendley MK, Singh P, Garg AK, Tiwari SP, Kumari K, Dutta GK. The studies on nutrient balances in crossbred cattle bulls fed chopped green sugarcane tops supplemented with some agro industrial by-products. Tropical Animal Health and Production 2009; 41: 943–949.

[9] Ortiz-Rubio MA, Ørskov ER, Milne J, Galina HMA. Effect of different sources of nitrogen on in situ degradability and feed intake of Zebu cattle fed sugarcane tops (*Saccharum officinarum*). Animal Feed Science and Technology 2007; 139: 143–158.

[10] Alcántara E, Aguilera A, Elliot R, Shimada A. Fermentation and utilization by lambs of sugarcane fresh and ensiled with and without NaOH. Animal Feed Science and Technology 1989; 23: 323–331.

[11] Bernardes TF, Silveira RN, Coan RM, Reis R, Moreira AL, Iturrino RPS. Fermentative characteristics and presence of yeast on raw or burnt ensiled sugar cane with additive. Reunião daSociedade Brasileira de Zootecnia, 39, 2002. Anais, Recife–Brazil (CD ROM).

[12] Nyakira BS, Tuitoek JK, Onjoro PA, Ambula MK. Determination of the nutritive value of sugar cane tops, mulberry leaves (*M. alba*) and calliandra (*C. calothyrsus*) as feed supplements for goats in Kenya. Journal of Animal Science Advances. 2015; 5: 1225–1233.

[13] Mthiyane DMN, Nsahlai IV, Bonsi MLK. The nutritional composition, fermentation characteristics, in sacco degradation and fungal pathogen dynamics of sugarcane tops ensiled with broiler litter with or without water. Animal Feed Science and Technology 2001; 94: 171–185.

[14] Salinas-Chavira J, Almaguer LJ, Aguilera-Aceves CE, Zinn RA, Mellado M, Ruiz-Barrera O. Effect of substitution of sorghum stover with sugarcane top silage on ruminal dry matter degradability of diets and growth performance of feedlot hair lambs. Small Ruminant Research 2013; 112: 73–77.

[15] Arribas L. Agriculture as a problem: worldwide citrus export. Agrícola Vergel, Fruticultura, Horticultura, Floricultura, Year XX, Num., 230, 2001; 56 p.

[16] Bampidis VA, Robinson PH. Citrus by-products as ruminant feeds: a review. Animal Feed Science and Technology 2006; 128: 175–217.

[17] Spreen TH. The citrus industries of the United States and Mexico after Nafta. Revista Chapingo Serie Horticultura 2000; 6: 145–152.

[18] Aguilera A, Perez-Gil F, Grande D, de la Cruz I, Juarez J. Digestibility and fermentative characteristics of mango, lemon and corn stover silages with or without addition of molasses and urea. Small Ruminant Research 1997; 26: 87–91.

[19] Migwi PK, Gallagher JR, Van Barneveld RJ. The nutritive value of citrus pulp ensiled with wheat straw and poultry litter for sheep. Australian Journal of Experimental Agriculture 2001; 41: 1143–1148.

[20] Itavo LCV, dos Santos GT, Jobim CC, Voltolini TV, Bortolassi JR, Ferreira CCB. Conservation of fresh orange peel by ensilage process using additives. Revista Brasileira de Zootecnia 2000; 29: 1474–1484.

[21] Itavo LCV, dos Santos GT, Jobim CC, Voltolini TV, Faria KP, Ferreira CCB. Composition and apparent digestibility of orange peel silage additives. Revista Brasileira de Zootecnia 2000; 29: 1485–1490.

[22] Revuelta-Llano D, Mosquera-López D, Cuba-Mora F. Ensiling potential of orange fruit wastes (Citrus sinensis). Revista Ciencias Técnicas Agropecuarias 2008; 17: 41–44.

[23] Malla BA, Rastogi A, Sharma RK, Ishfaq A, Farooq J. Kinnow madarin (*Citrus nobilis* lour × *Citrus deliciosa* tenora) fruit waste silage as potential feed for small ruminants. Veterinary World 2015; 8: 19–23.

[24] Pagán S, Rodríguez AA, Valencia EM, Randel PF. Pineapple and citrus silage as potential feed for small ruminant diets: fermentation characteristics, intake, nutrient digestibility, and aerobic stability. Revista Colombiana De Ciencias Pecuarias 2013; 27: 37–46.

[25] Lozicki A, Koziorzebska A, Halik G, Dymnicka M, Arkuszewska E, Niemiec T, Bogdan J. Effect of ensiling pumpkin (*Cucurbita maxima* D.) with dried sugar beet pulp on the content of bioactive compounds in silage and its antioxidant potential. Animal Feed Science and Technology. 2015; 206: 108–113.

[26] Halik GD, Lozicki A, Koziorzębska A, Dymnicka M, Arkuszewska E. Effect of ensiling pumpkin Cucurbita maxima with the addition of inoculant or without it on chemical composition and quality of silages. Annals of Warsaw University of Life Sciences – SGGW, Animal Science. 2014; 53: 103–110.

[27] Medina LJB, Salinas-Chavira J, Martinez DR, Lerma-Doria CE. 1993. Silage of ripe pumpkins (*Cucurbita maxima*) with sorghum ratoons, urea and molasses in complete rations for sheep. Available from: http://agris.fao.org/agris-search/search.do?recordID=MX19950101212 [accessed: 2016-02-01].

[28] Medina LJB, Salinas-Chavira J, Lerma DEC, Martinez Delgadillo R, Yado-Puente R. 1995. In vivo digestibility of squash silage (*Cucurbita maxima*) with sorghum hay, urea and sugar cane molasses in integral diets for ovine. VIII Congreso nacional de produccion ovina. Memorias. Chapingo, Mexico, pp. 77–81.

[29] Hashemi A, Razzaghzadeh S. Investigation on the possibility of ensiling cucurbit (*Cucurbita pepo*) residues and determination of best silage formula. Journal of Animal and Veterinary Advances 2007; 6: 1450–1452.

[30] Razzaghzadeh S, Amini-jabalkandi J, Hashemi A. Effects of different levels of Pumpkin (*Cucurbita pepo*) residue silage replacement with forage part of ration on male buffalo calves fattening performance. Italian Journal of Animal Science 2007; 6 (Suppl. 2): 575–577.

[31] Anjos FR, Tivana L, Da-Cruz-Francisco J, Kagande SM. Cassava (Manihot esculenta crantz): an affordable energy source in dairy rations. Online Journal of Animal and Feed Research 2014; 4: 10–14.

[32] Maza AL, Vergara GO, Paternina DE. Chemical and organoleptic evaluation of maralfalfa silage (*Pennisetum* sp.) plus fresh cassava (*Manihot esculenta*) . Revista MVZ Córdoba. 2011; 16: 2528–2537.

[33] Saucedo GC, Gonzalez PB, Revah SM, Viniegra GG, Raimbault M. Effect of lactobacilli inoculation on cassava (manihot-esculenta) silage—fermentation pattern and kinetic-analysis. Journal of the Science of Food and Agriculture 1990; 50: 467–477.

[34] Meraz M, Shirai K, Larralde P, Revah S. Studies on the bacterial acidification process of cassava (Manihot-Esculenta). Journal of the Science of Food and Agriculture 1992; 60: 457–463.

[35] Goncalves JAG, Zambom MA, Fernandes T, Mesquita EE, Schimidt E, Javorski CR, Castagnara DD. Chemical composition and profile of the fermentation of cassava starch by-products silage. Bioscience Journal 2014; 30: 502–511.

[36] Maciel RP, Neiva JNM, Oliveira RC, de Araujo VL, Lobo RNB. Fermentative patterns and chemical composition of Elephant-Grass silages with cassava by-product. Revista Ciência Agronômica 2008; 39: 142–147.

[37] Zambom MA, Fernandes T, Schmidt EL, Goncalves JA, Pozza MS, Javorski CR, de Souza LC, Tinini RC. Silage of residue from the extraction of cassava starch in diets from lactating Holstein cows. Semina: Ciências Agrárias, Londrina 2015; 36: 1701–1712.

[38] Despal D, Lestad A, Permana IC, Hidayah P. 2012. Effect of age at harvest on whole Cassava (*Manihot esculenta*) silage qualities. Proceedings of the 15AAAP Animal Science Congress. Thailand, pp. 594–599.

[39] Arce J, Rojas A, Poore M. Effect of poultry litter addition on the nutritional and fermentative characteristics of cassava (*Manihot esculenta*) by-product silage. Agronomía Costarricense 2015; 39: 131–140.

[40] Gomez G, Valdivieso M. Cassava for animal feeding: effect of variety and plant age on production of leaves and roots. Animal Feed Science and Technology 1984; 11: 49–55.

[41] Man NV, Wiktorsson H. Effect of molasses on nutritional quality of Cassava and Gliricidia tops silage. Asian-Australian Journal of Animal Science 2002; 15: 1294–1299.

[42] Oni AO; Sowande OS, Oni OO, Aderinboye RY, Dele PA, Ojo VO A, Arigbede OM, Onwuka CFI. Effect of additives on fermentation of cassava leaf silage and ruminal fluid of West African Dwarf goats. Archivos de Zootecnia 2014; 63: 449–459.

Section 3

Nutritive Value of Silages

Chapter 6

Intake and Digestibility of Silages

Juliana Silva de Oliveira, Edson Mauro Santos and
Ana Paula Maia dos Santos

Additional information is available at the end of the chapter

http://dx.doi.org/10.5772/65280

Abstract

The intake of DM (DMI) is determinant for ingress of nutrients to cater to the requirements for animal maintenance and production, principally the intake of protein and energy. The end-products of fermentation can affect the intake of silages and influence animal performance, since some organic acids negatively influence the intake of silage and digestibility of nutrients. For example, acetic and butyric acid have large effects on the intake of silage. Ammonia also can negatively affect the intake of silages. The digestibility can be influenced by end-products of fermentation and change the characteristics of ensiled plants. The objective of this chapter is to explain how silage end-products of fermentation and changes in the structure of forage resulting from the ensiling process can affect the intake and digestibility of silages. Some control mechanisms of silage fermentation can be used to improve the intake and digestibility of silage. Biological or chemical additives may contribute to the increased intake of silage and improve digestibility. Appropriate management techniques can influence the result.

Keywords: acetic acid, ammonia, animal nutrition, butyric acid, forage ensiling

1. Introduction

Ensilage is the method of forage conservation based on conversion of water-soluble carbohydrates in organic acids by the activity of lactic acid bacteria, which reduces the pH and preserve the fresh forage [1].

The ensiling process show advantages such as conservation of large quantities of forage in short time and forage conservation is less weather dependent. However, a disadvantage of the ensiling process is the relative reduction of feeding value of the silage when compared to the original crop [2]. Although, correct management of silages and use of additives may stimulate the intake of silages and improve the digestibility of silages.

The forages may have changes in the nutritive value due to the procedures during production, conservation and post-opening management and biochemical and microbiology phenomena [3]. Besides the conversion of carbohydrates in organic acids occurs by the partial breakdown of proteins, which gives the non-protein structures. These changes depend on the interaction between microorganisms on the material to be ensiled and amount and type of the substrate [4].

The silage quality affects the intake and digestibility of ruminants. Basically, the main factors that can interfere with the fermentation of the silage is the dry matter (DM) content, water-soluble carbohydrate (WSC) concentration and microorganism populations present in the forage. Forage with low DM and WSC concentrations may show undesirable fermentation and forage with excessive WSC content may generate acidic silages, which reduces silage intake.

Some end-products of fermentation, such as acetic and butyric acids and ammonia, are associated with the decrease in the intake of silages. Poorly fermented silages have large concentrations of undesirable compounds that explain the low silage intake.

Some changes resulting from the ensiling process influence the digestibility of silages. High concentration of ammonia, for example, can interfere in digestibility of the silage, and enter the N recycling cycle and increase the animal energy costs.

The objective of this chapter is to explain how end-products of fermentation of the silage and changes in the structure of forage resulting from the ensiling process can affect the intake and digestibility of silages.

2. Effect of silage quality on the intake and digestibility of ruminants

The feeding value of silage is mainly determined by intake and digestibility of silage [5]. The silage quality and availability of nutrients influence animal performance [6].

The intake of silage is generally lower than the intake of fresh forage [1] because the presence of toxic substances produced during the fermentation as amines; also due to the high concentration of organic acids and decrease in the water soluble carbohydrate content which lower availability of energy for the growth of microorganisms in the rumen [7]. However, we should question the validity of this conclusion [2].

Proper management in the ensilage can result in well-preserved silage and result in a similar intake of fresh forage. In addition, the use of additive may increase the silage intake values. Balieiro Neto et al. [8] evaluated the intake of sugarcane *in natura* and silage and observed higher values to sugarcane silage intake (0.720 kg/day) than to sugarcane *in natura* intake (0.657 kg/day). The silages were additive with 0.5% of calcium oxide (fresh matter basis).

The fermentative profile and the vegetable species available can influence the silage intake. Due to the fermentative process, many changes occur in chemical characteristics of forage. The

http://dx.doi.org/10.5772/65280

organic acid concentration is variable and it is influenced by management of the ensilage process, use of additive and principally by forage characteristics. This variation is recognized and search object of researchers worldwide (**Table 1**).

Crop	Ensiling characteristics								Author
	pH	LA (g/kg DM)	AA (g/kg DM)	PA (g/kg DM)	BA (g/kg DM)	Ethanol (g/kg DM)	NH_3-N (mM/g DM)	WSC (g/kg DM)	
Corn silage (*Zea mays* L.)	3.7	38.6	14.3	1.4	2.0	3.5	10.1	17.5	Hassanat et al. [12]
Sorghum silage (*Sorghum bicolor* L. Moench)	3.66	61.5	16.7	0.1	0.4	18	6.68	–	Santos [13]
Pearl millet silage (*Pennisetum glaucum* LR)	3.9	63	14.4	2.8	0.71	–	48	–	Dos Santos et al. [14]
Calliandra silage (*Calliandra calothyrsus*)	4.0	20.4	5.85	0.15	0.38	–	43.5	17.99	Ridwan et al. [15]
Alfafa silage (*Medicago sativa*)	4.2	74.2	26.1	2.4	0.9	2.2	27.7	18.3	Hassanat et al. [12]
Napiergrass silage (*Pennisetum purpureum* Schum)	4.1	37.6	12.7	0.1	12.1	–	100.1*	21.2	Rong et al. [16]
Cactus palm silage (*Opuntia ficus indica*)	3.81	80.2	22.5	8.1	05	–	9.0	20.4	Nogueira [17]

*NT.

Table 1. Fermentative characteristics of silages.

Restle et al. [9] evaluating the performance of feedlot calves receiving grass silage (*Brachiaria plantaginea*) and corn and sorghum silages, found that corn and sorghum silage promoted higher intake and better performance than the animals fed grass silage. Although the authors justify the higher dry matter intake of animals fed with corn and sorghum silage was due to an increasing difference in weight, which interfere directly in the intake values, the result may have also occurred because the fermentative characteristics and different end-products of fermentation concentration.

3. Factors that interfere on the silage intake

The dry matter intake (DMI) is determinant to ingress of nutrients to cater to the requirements for animal maintenance and production, principally the intake of protein and energy [10]. The DMI is the factor that affects the animal productive performance, since 60–90% variation in animal performance is associated with the metabolizable energy intake and only 10–40% with the diet digestibility [11].

The silage fermentative profile can influence the animal intake. In **Table 1**, the fermentative profile of some silages used in animal feed is described. The corn silage (*Zea mays* L.) shows a good fermentation process and result in adequate lactic acid production (38.6 g/kg DM), low acetic, propionic and butyric acid concentrations [12], which implies adequate dry matter intake.

The sorghum silage (*Sorghum bicolor* L. Moench) has similar fermentative characteristics of corn silage, but in some cases, the higher WSC content of forage can cause acid silage and increase the ethanol produce due to yeast activity [13]. The lactic acid production in sorghum ensilage is quick and pH may decrease below than desirable pH. Sorghum silage can show average values lactic acid of 61.5 g/kg DM and 18 g/kg ethanol concentration [13]. Ethanol may result in decrease in DMI, so excess ethanol is a negative point in these silages.

Another important forage to semiarid regions is the pear millet (*Pennisetum glaucum* LR). The fermentative profile of pear millet silage has high volatile fatty acids (VFA) products, with lactic acid 63 g/kg DM and higher propionic acid content (2.8 g/kg DM) [14] than corn and sorghum silages.

The forage species is an important factor for determining the fermentative profile and intake silage. Silages legumes, such calliandra (*Calliandra calothyrsus*), for example, has higher ammonia content (43.5 mM/g DM) [15] compared to corn silage. This large ammonia amount can influence the intake and digestibility of silages.

Alfalfa silage (*Medicago sativa*) have large amount of organic acid. Research evaluation of replacing effects of alfalfa silage for corn silage, Hassanat et al. [12] found different concentrations of organic acids into silages. The lactic acid content of alfalfa silage (74.2 g/kg DM) is higher than corn silage. However, other compounds such as acetic acid and ammonia can decrease intake silage. The high values found (acetic acid 26.1 g/kg DM and ammonia 27.7 mM/g DM) may have negatively influenced the silage intake by cows. The silage intake increased according to elevated levels of corn silage in the diet. Probably, the differences of the fermentative profile of silages alter the intake by the animal.

Grasses ensiling result generally in higher pH and lower values of lactic acid. Resistance to change in pH or buffer capacity is one of the main obstacles to the quality of silage. The rapid lowering of the pH is effective in reducing the activity of deleterious microorganisms to nutrient forage ensiled. Although the buffering substance content may hamper acidification of the silo environment, Rong et al. [16] observed pH values of 4.1 in Napiergrass (*Pennisetum purpureum* Schum) silage; still, high butyric (12.1 g/kg DM) and acetic acid concentrations (12.7 g/kg DM) and ammonia content (100 g/kg NT). Butyric acid can negatively influence the silage intake in ruminant animals [2].

In semiarid regions, the use of cactaceous in animal feeding is unexceptional. Nogueira [17] tested the cactus palm (*Opuntia ficus* indica) ensiling and found large amount of lactic acid (80.2 g/kg DM) and a high propionic acid content (8.1 g/kg DM). These acids have no relation

http://dx.doi.org/10.5772/65280

with the decrease in silage intake. Then it is possible that cactus palm silage may show a high animal intake index.

Chemical composition of forage pass by changes that alter the forage structure ensiling (**Table 2**). When ensiling, maize showed decrease in WSC (-59.5 g/kg DM), Neutral detergent fiber (NDF) (-25 g/kg DM), hemicellulose (-25 g/kg DM) and cellulose contents (-11g/kg DM), and increase on acid detergent lignin (+11 g/kg DM) [18].

Crop	**Chemical composition**									**Author**
	DM	**CP (g/kg DM)**	**WSC (g/kg DM)**	**NDF (g/kg DM)**	**ADF (g/kg DM)**	**ADL (g/kg DM)**	**HEM (g/kg DM)**	**CELL (g/kg DM)**	**ASH (g/kg DM)**	
Fresh maize	297.0	–	88.0	555.0	325.0	57.0	230.0	268.0	–	Filya and Sucu [18]
Maize silage	297.0	–	28.5	530.0	325.0	68.0	205.0	257.0	–	Filya and Sucu [18]
Fresh sorghum	361.0	108.0	67.0	608.0	359.0	65.0	–	–	46.0	Amer et al. [57]
Sorghum silage	352.0	116.0	18.0	609.0	361.0	43.0	–	–	50.0	Amer et al. [57]
Fresh pearl millet	230.8	116.8	–	595.0	321.4	–	273.6	279.8	–	Guimarães Jr. [58]
Pearl millet silage	241.9	112.5	–	486.8	288.8	–	198.1	256.2	–	Guimarães Jr. [58]
Fresh calliandra	450.5	212.1	–	551.8	488.8	215.7	63.0	299.8	–	Ridwan et al. [15]
Calliandra silage	465.4	202.2	17.9	538.4	448.8	134.1	89.6	307.0	–	Ridwan et al. [15]
Fresh Piatã grass	184.0	139.0	–	673.0	372.0	43.0	–	–	–	Costa et al. [59]
Piatã grass silage	265.5	97.8	–	632.5	405.8	43.0	–	–	–	Costa et al. [59]
Fresh *P. Purpureum*	312.5	45.2	–	746.2	510.1	382.0	236.2	73.5		Ridwan et al. [15]
P. Purpureum Silage	311.9	56.0	–	665.8	492.1	343.6	173.7	87.9		Ridwan et al. [15]

Table 2. Chemical composition of pre-ensiling of forage and silage.

Some forage after ensiling, as Pearl Millet and Calliandra, show increase in the DM content and decrease in CP concentration and fibrous fractions of forage (NDF, NDA, ADL, hemicellulose and cellulose). The reduction in fibrous fraction is positive to silage degradability, because it expands the activity area of rumen microorganisms and resealing energy.

Others forages, such as sorghum, may show higher ash content after ensiling. This can occurs due the biochemistry reactions of organic acids and salt formation. The exposure of the silage to air, making the anaerobic environment to aerobic, is one of the factors that influence the nutritional value of silages. In the presence of air, deleterious and opportunistic aerobic microorganisms can develop rapidly and degrade nutrients from silages.

Sugarcane and corn are forage susceptible to aerobic stability problems because the high lactic acid concentration and residual WSC promote an ideal environment for the development of deleterious microorganisms, such yeasts. In **Table 3**, it is observed that after exposure to air over time (8 and 9 days), there was a reduction in the WSC content of corn silage and an increased DM content [8]. The sugarcane silages showed an increase in the NDF and CP contents and reduced lignin (ADL) and non-fibrous carbohydrate (NFC) contents [19]. These changes in the chemical composition of silages can interfere on intake and digestibility of silages.

Crop	**Chemical composition**							
	DM (g/kg DM)	**CP (g/kg DM)**	**NDF (g/kg DM)**	**ADF (g/kg DM)**	**ADL (g/kg DM)**	**NFC (g/kg DM)**	**WSC (g/kg DM)**	**ASH (g/kg DM)**
Sugarcane	269.1	30.0	554.8	439.5	72.5	375.8	–	–
Sugarcane silage	238.2	31.9	633.3	479.2	83.7	282.0	–	–
Sugarcane silage (d3)	340.5	38.9	659.7	493.3	82.3	245.2	–	–
Sugarcane silage (d6)	359.2	37.8	709.7	440.4	69.2	195.8	–	–
Sugarcane silage (d9)	299.7	42.7	704.7	414.6	59.4	196.0	–	–
Maize	339	71	409	198	–	–	–	35
Maize silage	317	78	384	206	–	–	18	37
Maize silage (d0)	360	75	354	203	–	–	17	35
Maize silage (d2)	366	73	370	209	–	–	18	37
Maize silage (d4)	371	76	358	217	–	–	15	35
Maize silage (d6)	389	75	356	208	–	–	9	35
Maize silage (d8)	395	76	362	206	–	–	11	35

Source: Adapted from Balieiro Neto et al. [8] and Gerlach et al. [19].

Table 3. Changes in the chemical composition of fresh forage, silage at silo opening and silage exposure to air.

In ruminant animals, the intake is regulated by psychogenic, physiological or physical mechanisms [11]. The psychogenic mechanism is related to aspects of smell and palatability of the food [20].

Palatability is the property of a food that affects its taste or smell as perceived by animals with particular experiences under specified conditions. The palatability may be a basis to the silage intake problem [21]. The end-products of fermentation in silage can affect the animal intake through palatability [22]. In addition, the palatability depends on the animal species.

The regulation of intake in ruminants can occur through of humoral factors because the volatile fatty acids (VFA) have the ability to limit the intake. The intake varies with the energy requirements of the animal. The physiological mechanism can be observed when provided with high concentrate diets, as animals in confinement. As silage has a considerable content of organic acids due the fermentative process, the intake of silage tends to be lower than the original forage. Some silage fermentation products can reduce the intake of silage, such as acetic acid [23].

Animals with diet rich in forage prevails the intake limited by the physical capacity rumen. The NDF is the main fraction of diet that provides this effect because it is slow and incomplete digestion in gastrointestinal [11]. In this situation, repletion has a significant effect on animal capacity in DM intake. The physical distension of the reticulum-rumen is the main factor limiting the intake of fodder and many diets rich in fiber [14].

Fiber fraction is an important parameter to be considered for the animal intake, because it is negatively correlated with intake and digestibility [7]. Even with the lower fiber content when compared to hay, silage tends to be less intake by the animals. In the experiment evaluating the effects of fermentation on the intake and digestibility of silage [24], the authors found an average of 16% lower intake of silage compared to the intake of hay. They noted that this reduction in intake was due to the presence of fermentation end-products.

The major source of ingredients in dairy cow diets is the forage. Despite their important (economically and nutritionally), the forage has been a study object for a long time [12].

Crop	**Intake dry matter (kg/d)**	**Digestibility**			**Animal species**	**Author**
		IVD (%DM)	**TIVD (%DM)**	**AD (%DM)**		
Corn silage	22.80	–	–	71.30	Cow	Hassanat et al. [12]
Alfalfa silage	21.70	–	–	69.70	Cow	Hassanat et al. [12]
Sugarcane *in natura*	–	63.93	66.50	–	–	Balieiro Neto et al. [8]
Sugarcane silage	–	59.72	62.11	–	–	Balieiro Neto et al. [8]
Sorghum silage (SS)	5.91	–	–	48.32	Sheep	Simon et al. [35]
SS + 15% concentrate	7.09	–	–	61.96	Sheep	Simon et al. [35]
SS + 15% concentrate	7.81	–	–	68.12	Sheep	Simon et al. [35]
SS + 15% concentrate	7.98	–	–	69.77	Sheep	Simon et al. [35]

IVD = *in vitro* digestibility; TIVD = true *in vitro* digestibility; AD = apparent digestibility.

Table 4. Intake and digestibility of silage index.

Evaluating the effect of replacing alfalfa silage to corn silage on diet intake of dairy cows, among other parameters, Hassanat et al. [12] found values of 21.7 kg/day DMI alfalfa silage and 22.8 kg/day DMI corn silage (**Table 4**). The difference on intake silage in these silages may occur due chemical composition and fermentative profile. As mentioned before, butyric and acetic acid are associated with decreased intake silages. These acids were found in greater quantity in alfalfa silage than in corn silage.

3.1. End-products of fermentation

The ensilage is a complex process and it yields a variety of compounds and forage quality is a term used to refer to the nutritional value of plant in interaction with the animal intake and performance potential. About silage, the animal response is dependent on its fermentative profile that affects the food structure, nutrient concentration and intake [3].

The quality of the silage is influenced by factors such as plant species, indigenous microbiota, crop management, cutting and ensilage procedure, as well as environmental factors and storage. The variation of silage quality can influence the intake by animals. The composition and concentration of fermentation end-products of the silage are variable, most commonly found fermentation produces lactic acid, however, other types of fermentations occur and may decline the nutritional quality of the silage [4]. The fermentation quality should be included in assessment of the DMI potential of grass silages [24].

Although intensively discussed, there is no agreement on the indices of fermentation quality when evaluating the dry matter intake of the silage [5, 24], however, some factors may be identified for reducing the intake of silage.

Researchers evaluated the production of fermentative compounds in silages and they found 13 esters, 5 aldehydes, 3 alcohols and 1 sulfide. They observed that the increase in ammonia, acetate and propionate levels, as well as decrease in the WSC content, decrease the intake [22].

Esters are volatile compounds that may take effect on silage flavor reducing the DMI, principally acetate [19]. Other compounds also may influence the DMI, as propionic acid and biogenic amine. According to some studies it is improbable that low propionic acid concentration directly influence the DMI, while the biogenic amine is naturally present in silages and reduce the intake from palatability or by influencing the nitrogen metabolism [5].

An indicator that can be related to the DMI is pH. Researchers evaluating changes in the fermentation of corn silage exposed to air [16] found a positive correlation between pH and DMI, if pH is relatively high, and a greater intake of silage. This is justified because of the absence of excessive organic acids or ammonia-N fermentation [5, 19].

The fermentation process of the silage not just can generate products that inhibit the animal intake; the reactions resulting from the silo opening procedure can also promote reduction of silage intake. The aerobic deterioration is a significant problem that affects the yield and quality of silage [25]. It is caused by the activity of bacteria, yeasts and molds that can compromise the final nutritional value of the silage, changing the volatile and depressed intake [19].

http://dx.doi.org/10.5772/65280

In a study on the effect of aerobic deterioration of silage on goat intake, the authors found intense degradation of lactic (by yeasts) and acetic acids, decrease in WSC after air exposure. After eight days, it was more than half with reductions ranging between 29 and 79% in comparison to fresh silages. Some end-products of fermentation (ethyl lactate, ethanol) were negatively related to silage intake. However, correlation coefficients were weak. Concentrations of acetic acid and ethanol were negatively correlated with DMI, but the authors justified that lower DMI is due to the greater concentration of acetic acid in fresh silages which compensate for improved aerobic stability and smaller decline in DMI is a consequence of aerobic deterioration [19].

3.1.1. *Lactic acid*

The lactic acid is an organic acid produced by conversion of soluble carbohydrates by lactic acid bacteria. The lactic acid content should be at least 65–70% of the total silage acids in good silage [26].

The lactic acid concentration in silages may to decrease the efficiency of microbial protein synthesis in the rumen whenever the values are high [27].

3.1.2. *Acetic acid*

Extremely wet silages or slow silo filling can result in silages with high concentrations of acetic acid (>3–4% of DM). Acetic acid concentrations are also related principally to long-acting of enterobacteria and heterofermentative bacteria [26].

The acetic acid content negatively relates to the intake of silage [28], therefore low levels of acetic acid are desirable in silages [29]. Though it may present a negative aspect to intake, silage inoculants specific with the largest production of acetic acid did not show a reduction in animal intake [26].

In assay sheep fed with silage, the authors [21] found decrease in the intake of silage when adding acetic acid. The reduction in intake is justified due to the taste and odor of silage. In other studies, DMI was negatively correlated with acetic acid [19].

3.1.3. *Butyric acid*

Concentrations of butyric acid are negatively correlated with digestibility [24]. A high concentration of butyric acid (>0.5% of DM) indicates that the silage has a poor fermentation, clostridia fermentation. Silages high in butyric acid have a low nutritive value and many of the soluble nutrients have been degraded. They may contain compounds such as amines that have sometimes shown adversely affect animal performance, also the intake [26].

The butyric acid content reflects clostridia activity on ensiled mass with a deleterious effect on quality and reduction of silage palatability [29].

3.1.4. Ammonia

The ammonia-N is often associated with the decrease in the intake of silage because of their presence in poorly fermented silage or clostridia. Some other products resulting of degradation of amino acids also can decrease the intake of silage [21].

The proteolysis by plant and microbial enzymes may lower the nutritive value of ensiling forage by degrading the forage protein fraction into peptides, amines, free amino acids and ammonia. This permit proteolytic bacteria ferment peptides and amino acids converting them into a diversity of organic acids, CO_2, ammonia and amines, products that decrease the voluntary intake of silage [30]. Generally, the ammonia concentration is used as an indicator of protein degradation in silage [5].

Although microorganisms as enterobacteria have low proteolytic activity, it can deaminate and decarboxylate some amino acids contributing to the formation of ammonia and biogenic amines in silage, which have a negative effect on silage palatability and intake in ruminants [31].

Ammonia concentrations are negatively related to the intake of silage. In grass silage, high moisture favors the butyric fermentation and release of ammonia, which negatively affect the intake of silage by animals [1]. According to Huhtanen et al. [5] index, ammonia concentrations greater than 50 g/kg N predict decrease in silage DMI.

4. Factors that affect silage digestibility

Digestion is a process of conversion of food macromolecules into simple compounds that can be absorb into the gastrointestinal tract [10].

Concentration of ammonia-N in rumen is indispensable for microbial growth since it is associated with the energy source, and it is related to soluble protein of diet and to N retention of the animal. High ammonia concentrations may occur when excess protein in the diet is degraded in the rumen or a low concentration of carbohydrates is degraded in the rumen, which can cause changes in rumen pH changing the microbial activity and its functions in the digestive process [10].

The digestibility of ruminants is associated with the characteristics of food and the animal. The relationship between intake and digestibility, in which the increase in digestibility leads to increased intake, is influenced by forage residence time in the rumen [32].

Some factors, such as the proper processing of forage for silage, contribute to improve the digestibility of the final product. Researchers evaluated the effect of the length of the whole plant corn on intake, digestibility and production of milk [33], they found positive results regarding the effect of whole plant processing for corn silage with increased body weight, increased DM intake, greater starch and fiber digestibility. Still, the estimated average increase in starch digestibility in the ensiled plants was 4.2% above the initial herbage unprocessed digestibility.

The yield of fermentation end-products in silage is variable, depending primarily on the amount of substrates and microbial flora. Some silage may contain up to 200 g/kg DM of fermentation end-products, especially lactic acid and VFA, which provide low energy for the rumen microorganisms [34].

Other factors such as exposure to air and use of additives can influence silage digestibility. The fermentation type interferes in the result of silage intake and digestibility. Corn and alfalfa silages have different digestibilities (**Table 4**). Compared to fresh forage, digestibility of silages is lower [8], but this can be modified.

The use of additives as concentrate ration can increase intake and digestibility of silages. The total mixture ration is an efficient technique and may increase intake and digestibility of silages, obtaining considerable increases in apparent digestibility [35].

4.1. Changes in the fermentation process that affects silage digestibility

Degradability of silage is positively correlated with WSC and LA [24].

The fresh forage has approximately 75–90% of the total nitrogen present in the protein form [29], the rest called non-protein nitrogen comprises free amino acids and amides, and ammonia with concentration less than 1% of total nitrogen. During the fermentative process of silage, part of nitrogen fraction is degraded to soluble fractions as peptides, amino acids and ammonia, which are rapidly degraded in the rumen with low microbial synthesis efficiency and results in inappropriate protein post-rumen flow [36].

According to the research of Mckersie, in 1985 [29], compounds resulting from proteolysis and degradation of amino acids formed during fermentation of silage can inhibit the intake and have low utilization efficiency of the microorganisms present in the rumen [29]. The concentration of ammonia in good silages should be low, not to influence the silage intake negatively [7].

During the ensiling process the breakdown of hemicelluloses occurs to provide additional substrate for the fermentation, because concentrations of NDF in the silages are lower than the original herbage. The degradation of hemicellulose also can occur through hydrolysis by organic acids or action additives [4].

Compared with the herbage, concentrations of NDF can be altered by breaking the nitrogen bound to NDF, but an increase in the concentrations of NDF and ADL in silage may occur due to DM losses or effluent losses of soluble nutrients [24]. Concentration of ADL increases also due to synthesis of Maillard polymers [7], which may present positive correlation with ADIN. The changes in the fiber fractions attributable to the fermentative process of silage could influence digestibility [24].

Researchers evaluating different proportions of sorghum silage in diet of beef cattle compared to Tifton grass pre-dry, found an increase in dry matter, organic matter and total carbohydrate digestibility on adding a higher proportion of sorghum silage to diet. The authors justified that the increase of digestibility occurs due the lower NDF proportion and greater TDN (total digestible nutrients) which has rapid and complete availability in the gastrointestinal tract [10].

The exposure to air of silages affects the silage digestibility. The effects of air on silage can reduce the digestibility (**Table 5**). Sugarcane silage show lower digestibility than fresh sugarcane, and after 3, 6 and 9 days of exposure to air, sugarcane silage has reduced 7.20% of *in vitro* digestibility (IVD) and 2.7% true *in vitro* digestibility (TIVD). This reduction can be avoided or minimized by adequate ensiling management procedures and storage of silage, in addition, to the use of additives.

Crop	**Digestibility**	
	IVD (%DM)	**TIVD (%DM)**
Sugarcane	63.9	66.5
Sugarcane silage	59.7	62.1
Sugarcane silage (d3)	57.5	58.7
Sugarcane silage (d6)	54.5	58.3
Sugarcane silage (d9)	55.4	59.4

Source: Adapted from Balieiro Neto et al. [8].

Table 5. Effect of exposure to air on silage digestibility.

5. Alternatives to improve intake and digestibility of silages

5.1. Use of biological additives

The use of inoculants, especially lactic acid bacteria (LAB), is in an attempt to improve the efficiency of preserving the nutritional quality of the forage. In a review of experiments with inoculants, researchers found positive results for improving the feed intake, fed efficiency and milk production by about a third of the studies review; it is justifying the use of inoculants on silage also the effect on animal performance [34].

Some studies suggest a possible effect of LAB probiotics, although the mechanisms are unclear. Probiotics is a live microorganism in the food supplement that beneficially affects the host animal by improving intestinal balance [37]. One hypothesis is that specific strains of LAB interact with microorganisms of the rumen improving their function and animal performance [38–40]. Researchers found that LAB from silage inoculants could survive in rumen fluid for at least 96 hours, which would allow the probiotic activity [41].

Although the effects of LAB inoculant are not well studied, there may be still, the action of a type of bacteriocin that limits the bacterial activity, which can inhibit or harm the microorganism in the rumen [39, 40]. Bacteriocins are biologically active proteins produced by LAB that are active against other bacteria, mainly grampositive bacteria as *Listeria monocytogenes* [42]. In an experiment, Amado et al. [42] observed that bacteriocin producing strain inhibits the activity of other undesirable microorganisms in silage.

Recent studies demonstrate that some heterofermentative bacteria [*Lactobacillus buchneri*, for example] produce ferulate-esterase, enzyme that increases the degradation of the cell wall. This enzyme release considerable soluble carbohydrates for fermentation or for use by rumen bacteria [43].

An enzyme-bacterial inoculant acts in two forms in silage: whereas bacterial inoculants improve fermentation profile and increase lactic acid bacteria population, enzyme inoculants act on the cell wall and the available higher quantity of soluble compounds, with improvement in silage digestibility [44].

Researchers study the effect of inoculants on silage, rumen function and digestibility. They found improvements in DM and NDF digestibility after 24 hours of incubation [38]. Others studies also found higher DM and NDF digestibility in inoculated corn silage than untreated silage [39].

Although there are some positive results, in the experiment realized by Fugita et al. [42], the addition of enzyme-bacteria inoculants do not significantly influence nutrient intake, performance and carcass characteristics of feedlot finished crossbred bulls.

The use of microbial additives in sorghum silage resulted in positive responses to the hemicellulose content and value on *in vitro* DM digestibility. The lower hemicellulose content of the silage treated compared to control may result from the action of enzymes associated with bacteria, and the greater IVDMD found may reflect the enzymatic hydrolysis effect [45]. However, it has been reported that the effects of LAB inoculants on fiber degradation are not consistent [18] as LAB cannot use fiber as an energy source [46]. The hemicellulose degradation by LAB inoculation is inhibited in lower environmental temperature, requiring optimum temperature for its activity [47].

5.2. Use of chemical additive

Additives in silage can affect the DM intake and intervene in the nutritive value of silage, as digestibility of nutrients. Chemical additives are substances that act in the control of biochemical reactions of silage. The inhibitor additives function without distinction in all processes in the silage acting on undesirable microorganisms and fermentations, as the secondary proteolysis or aerobic growth. Among the main additives chemical inhibitors there are urea [48], propionic and formic acid.

Urea is an additive that contains between 42 and 45% of nitrogen [48], commonly used in fodder ammonization due to ease of application, not a pollutant, but as a source of non-protein nitrogen, reduce the fibrous portion of forage (NDF), favor the partial solubilization of hemicelluloses, influence the increase in intake and digestibility of silage [49]. According to the classification McDonald et al. [4], urea is also a nutrient additive because it improves the nutritive value of silage.

Researchers showed the increase in the protein content of silage as result of high recovery of nitrogen applied and may reach up 77% recovery [50]. Nitrogen recovery is a positive feature of urea from both the nutritional and economical aspect. Urea also acts beneficially in the fibrous portion of the ensiled forage. Two main processes occurring in ammoniated forage mass with urea: ureolysis and ammoniolysis.

The ureolysis process is an enzymatic reaction that release ammonia through hydrolysis of urea. The ureolytic bacteria produces urease [an enzyme catalyst present in plants], that acts in the presence of moisture hydrolyzing the urea and producing two ammonia molecules [which acts directly on the cell wall of forage] and one carbon dioxide [48, 51].

From the urea hydrolysis occur chemical reaction ammoniolysis between the ammonia and the ester bonds existing between chains of hemicelluloses and between groups of carbohydrates or carbohydrate molecules and lignin, resulting in formation of an amide [52]. The ammoniolysis cause lysis on bonds between the structural carbohydrates releasing and increasing the contact surface to the rumen microorganisms [53].

In addition, there is another important factor to consider, ammonia has a high affinity for water resulting in the formation of weak base, ammonium hydroxide [NH_4OH]. The high affinity of ammonia to water promotes expansion and rupture of the cell wall components of tissues of forage treated with urea. Through specific studies using electron microscopy, change of cell wall can be seen [49].

Another chemical additive, propionic acid is used as antifungal agent able of preserve forage for much time. It inhibits undesirable microorganisms and improves the aerobic stability of silages [54].

In an experiment test, Chen et al. [54] evaluated the effects and propionic acid applied on the fermentation quality and aerobic stability of total mixed ration silage (TMR) prepared with whole-plant corn in Tibet. They applied 0.4% propionic acid on a fresh matter basis of TMR, and found higher WSC concentration (88.92 g/kg DM) and decrease in butyric acid content (0.04 g/kg DM) in TMR after 45 days of ensiling, comparative to no additive TMR (WSC = 39.99 g/kg DM; butyric acid content = 0.19 g/kg DM). In aerobic stability assay, TMR silage with propionic acid showed low pH, higher WSC concentration and lower ammonia content than no additive TMR silage after 12 days of air exposure (**Table 6**).

	Control TMR silage				**TMR silage with 0.4% propionic acid**			
	0	**6**	**9**	**12**	**0**	**6**	**9**	**12**
pH	3.90	4.28	5.1	7.07	3.89	3.88	3.87	3.75
LA	86.53	66.95	38.63	15.49	65.44	83.34	83.42	67.09
WSC	39.99	29.16	34.64	29.56	88.92	78.23	72.23	54.23
NH_3-N	52.83	51.87	55.38	65.59	42.48	41.94	51.72	57.36

Source: Adapted from Chen et al. [54].

Table 6. Chemical characteristics of total mixture ration silages (TMR) at opening silo and after exposure to air.

The inhibition of undesirable microorganisms in silage (able to realize proteolysis) reduced the adverse compound formation. The decrease in ammonia and butyric acid in TMR silage is desirable because these compounds may affect food intake in the ruminants.

Besides propionic acid, formic acid is an inhibitor of undesirable fermentation. Selwet [55] evaluated the effects of different levels of mixtures of formic and propionic acid on changes in the chemical composition and on aerobic stability of maize silages exposed to air during the process of feeding to animals.

The results showed that the inclusion in maize silages of the propionic and formic acid mixture reduced undesirable microorganisms and positively influenced the changes in silage chemical composition. Silages treated with acids were characterized by higher dry matter, WSC and crude protein concentration, which could have been associated with the smaller losses of nutrients due the limitation of development of some groups of microorganisms [55].

Concentrations of acetic acid in additive silages were also decrease. The author concluded that this result is a favorable phenomenon because high concentration may limit feed intake by animals.

In an experiment, Kung et al. [56] tested different mixtures of preserving agents such as acetic and propionic acid and ammonia on the intake and digestibility of lactating cows fed TMR silages. The TMR were composed of alfalfa silage (27%), corn silage (43%) and pelleted concentrate (30%), and additives.

There was no significant difference between the dry matter intake, daily milk yield, fat and milk protein in dairy cows fed on untreated TMR and TMR treated with chemical additive after exposure to air. Although this study found no difference between the performances of dairy cows, other tests reported that feed intake by sheep was negatively affected after silage exposure to air for 5 days, when compared with fresh corn silage [56].

The use or not of a chemical or biological additive does not dispense the necessary care during the fermentative process of ensiling, because the quality of silage is directly related to species of plant, sol fertility, cultural tracts, ensiling point, compaction and sealing of silo, since only the additive does not match a considerable increase in silage quality produced [48].

6. Final considerations

The ensiling process is complex and yields a variety of end-products of fermentation. These products can influence directly and indirectly the intake and digestibility of silage. Some control mechanisms of this fermentation can be of use for improvement on intake and digestibility.

Biological or chemical nature, additives may contribute to the increased intake of silage, as well as improve digestibility. To choose the ideal additive, it is necessary to understand the factors that limit the intake and digestibility of food.

The exposure of silage to air is an inevitable phase and which may compromise the nutritional value of the silage when realized incorrectly. Appropriate management techniques can influence the result.

Author details

Juliana Silva de Oliveira[1*], Edson Mauro Santos[1] and Ana Paula Maia dos Santos[2]

*Address all correspondence to: oliveirajs@yahoo.com.br

1 Federal University of Paraiba, João Pessoa, Paraíba, Brazil

2 State University of Alagoas, Alagoas, Brazil

References

[1] Santos, E.M.; Zanine, A.M.; Oliveira, J.S. Tropical Grass silages. Revista Eletrónica de Veterinaria REDVET, v. VII, n. 07, pp. 1–16, 2006. DOI: 10.5747/ca.2006.v02.n1.a21

[2] Charmley, E. Towards improved silage quality – a review. Canadian Journal of Animal Science, v. 81, pp. 157–168, 2001.

[3] Jobim, C.C.; Nussio, L.G.; Reis, R.A.; Schmidt, P. Methodological advances in evaluation of conserve forage quality. Revista Brasileira de Zootecnia, v. 36, pp. 101–119, 2007. ISSN *on-line*: 1806-9290

[4] McDonald, P.; Henderson, A.R.; Heron, S. The biochemistry of silage. Marlow: Chalcombe. 2. ed. 1991. 340p.

[5] Huhtanen, P.; Khalili, H.; Nousiainen, J.I.; Rinne, M.; Jaakkola, S.; Heikkila, T.; Nousiainen, J. Prediction of the relative intake potential of grass silage by dairy cows. Livestock Production Science, v. 73, p. 111–130, 2002. S0301-6226[01]00279-2

[6] Mizubuti, I.Y.; Ribeiro, E.L.A.; Rocha, M.A.; Silva, L.D.F.; Pinto, A.P.P.; Fernandes, W.C.; Rolim, M.A. Intake and apparent digestibility of corn (zea mays l.), sorghum (sorghum bicolor [L] moench) and sunflower silages (helianthus annuus l.). Revista Brasileira de Zootecnia, v. 31, n. 1, p. 267–272, 2002.

[7] Van Soest, P.J. Nutritional ecology of the ruminant. Ithaca: Cornell University Press. 2. ed. 1994. 476p. ISBN: 0-8014-2772-X

[8] Balieiro Neto, G.; Ferrari Junior, E.; Nogueira, J.R.; Possenti, R.; Paulino, V.T.; Bueno, M.S. Fermentative losses, chemical composition, aerobic stability and apparent digestibility of sugar cane with chemical and microbial additive. Pesquisa Agropecuária Brasileira, v. 44, n. 6, pp. 621–630, 2009.

[9] Restle, J.; Neumann, M.; Brondania, I.L.; Gonçalves, J.M.; Pellegrinis, L.G. Avaliation of Papuã Grass silage [Brachiaria plantaginea] in performance of confined calf cattle. Ciência Rural. V. 33, n. 4, p. 749–756, 2003.

[10] Pereira, O.G.; Souza, V.G.; Valadares Filho, S.C.; Ribeiro, K.G.; Pereira, D.H.; Cecon, P.R. Intake, digestibility and ruminal parameters in beef cattle fed with sorghum silage and 85 Tifton grass pre-dried diets. Revista Brasileira de Zootecnia, v. 36, n. 6, pp. 2143–2151, 2007 ISSN *on-line*: 1806-9290

[11] Cabral, L.S.; Santos, J.W.; Zeivoudakis, J.T.; Abreu, J.G.; Souza, A.L.; Rodrigues, R.C. Intake and fed efficience in lambs. Revista Brasileira de Saúde e Produção Animal, v. 9, n. 4, pp. 703–714, 2008. ISSN: 1519 9940

[12] Hassanat, F.; Gervais, R.; Julien, C.; Massé, D.I.; Lettat, A.; Chouinard, P.Y.; Petit, H.V.; Benchaar, C. Replacing alfalfa silage with corn silage in dairy cow diets: effects on enteric methane production, ruminal fermentation, digestion, n balance, and milk production. Journal Dairy Science, v. 96, pp. 4553–4567, 2013.

[13] Santos, A.P.M.S. BRS ponta negra sorghum silages with urea. [Dissertação] Universidade Federal da Paraíba, Brazil. 57 p. 2014.

[14] Dos Santos, R.D.; Neves, A.L.A.; Pereira, L.G.R.; Sollenberger, L.E.; Rodrigues, J.A.S.; Tabosa, J.N.; Verneque, R.S.; Oliveira, G.F.; Jayme, D.G.; Gonçalves, L.C. Agronomic traits, ensilability and nutritive value of five pearl millet. Journal of Agricultural Science, v. 154, pp. 165–173, 2016.

[15] Ridwan, R.; Rusmana, I.; Widyastuti, Y.; Wiryawan, K.G.; Prasetya, B.; Sakamoto, M.; Ohkuma, M. Fermentation characteristics and microbial diversity of tropical grass-legumes silages. Asian Australian Journal Animal Science, v 28, n. 4, pp. 511–518, 2015.

[16] Rong, H.; Yu, C.; Li, Z.; Shimojo, M.; Shao, T. Evaluation of fermentation dynamics and structural carbohydrate degradation of napiergrass ensiled with additives of urea and molasses. Pakistan Veterinary Journal, v. 33, n 3, pp. 374–377, 2013.

[17] Nogueira, M.S. Fermentative profile and chemical composition of cactus palm silage with urea and wheat bran. [Dissertação] Universidade Federal da Paraíba, Brazil. 63 p. 2015.

[18] Filya, I.; Sucu, E.; The effects of lactic acid bacteria on the fermentation, aerobic stability and nutritive value of maize silage. Grass and Forage Science, v. 65, pp. 446–455, 2010.

[19] Gerlach, K.; Rob, F.; Weib, K.; Buscher, W.; Sudekum, K. Changes in maize silage fermentation products during aerobic deterioration and effects on dry matter intake by goats. Agricultural and Food Science, v. 22, p. 168–181, 2013.

[20] Mertens, D.R. Regulation of forage intake. In: Forage quality, evaluation and utilization. Wisconsin, 1994. WI 53706.

[21] Buchanan-Smith, J.G. As investigation into palatibility as a factor responsible for reduced intake of silage by sheep. Animal Production, v. 50, p. 253–260, 1990. DOI: 10.1017/S0003356100004700

[22] Krizsan, S.J.; Westad, F.; Adnoy, T.; Odden, E.; Aakre, S.E.; Randby, A.T. Effect of volatile compounds in grass silage on voluntary intake by growing cattle. Animal, v. 1, p. 283–292, 2007. DOI: 10.1017/S1751731107683773

[23] Alves, A.A.; Salles, R.O.; Azevedo, P.M.M.R.; Azevedo, A.R. Factors that affect the food intake by ruminants: a review. Rev. Cient. Prod. Anim., v. 3, n. 2, p. 62–72, 2001.

[24] Krizsan, S.J.; Randby, A.T. The effect of fermentation quality on the voluntary intake of grass silage by growing cattle fed silage as the sole feed. Journal Animal Science, v. 85, p. 984–996, 2007. DOI:10.2527/jas.2005-587

[25] Tabacco, E.; Piano, S.; Cavallarin, L.; Bernardes, T.F.; Borreani, G. Clostridia spore formation during aerobic deterioration of maize and sorghum silages as influenced by Lactobacillus buchneri and lactobacillus plantarum inoculants. Journal of Applied Microbiology, v. 107, pp. 1632–1641, 2009. DOI: 10.1111/j.1365-2672.2009.04344.x

[26] Kung, L.; Shaver, R. Interpretation and use of silage fermentation analysis reports. University of Wisconsin Board of Regents, 2001.

[27] Jaakkola, S.; Huhtanen, P. The effect of lactic acid on the microbial protein synthesis in the rumen of cattle. Asian-Australasian Journal of Animal Science, v. 2, n. 3, p. 398–399, 1989. SF 00710.

[28] McDonald, P. The biochemistry of silage. John Wiley & Sons, Ltd., 1981. 226 p. ISBN: 0-471-27965-X

[29] Tomich, T.R.; Gonçalves, L.C.; Tomich, R.G.P.; Rodrigues, J.A.S.; Borges, I.; Rodriguez, N.M. Chemical characteristics and in vitro digestibility of silage sunflower. Revista Brasileira de Zootecnia, v. 33, n. 6, p. 1672–1682, 2004.

[30] Senger, C.C.D.; Mühlbach, P.R.F.; Sanchez, L.M.B.; Netto, D.P.; Lima, L.D. Chemical composition and in vitro digestibility of maize silages with different maturities and packing densities. Ciência Rural, v.35, n. 6, p. 1393–1399, 2005. ISSN 0103-8478

[31] Driehuis, F.; Oude Elferink, S.J.W.H. The impact of the quality of silage on animal health and food safety: a review. Veterinary Quarterly, v. 22, n. 4, pp. 212–216, 2000. DOI: 10.1080/01652176.2000.9695061

[32] Nadeau, E.M.G.; Russell, J.R.; Buxton, D.R. Intake, digestibility and composition of orchardgrass and alfafa silages treated with cellulase, inoculant and formic acid fed to lambs. Journal Dairy Science, v. 78, p. 2980–2989, 2000.

[33] Bal, M.A.; Shaver, R.D.; Jirovec, A.C.; Shinners, K.J.; Coors, J.G. Crop processing and chop length of corn silage: effects on intake, digestion and milk production by dairy cows. Journal Dairy Science, v. 83, p. 1264–1273, 2000.

[34] Rinne, M.; Nousiainen, J.; Huhtanen, P. Effects of silage protein degradability and fermentation acids on metabolizable protein concentration: a meta-analysis of dairy cow production experiments. Journal Dairy Science, v. 92, pp. 1633–1642, 2009. DOI: 10.3168/jds.2008-1429

[35] Simon, J.E.; Lourenço Júnior, J.B.; Ferreira, G.D.G.; Santos, N.F.A.; Nahum, B.S.; Monteiro, E.M.M. Intake and digestibility ofsorghum silage as food supplement alternative to orient amazônia ruminants. Amazônia: Ci & Desenv., v. 4, n.8, pp. 103–119, 2009.

[36] Nussio, L.G.; Paziani, S.F.; Nussio, C.M.B. Ensiling tropical grass. In: Annual Meeting of the Brazilian Society of Animal Science. Anais… Recife: Brazilian Society of Animal Science, p. 60–83. 2002.

[37] Weinberg, Z.G.; Shatz, O.; Chen, Y.; Yosef, E.; Nikbahat, M.; Ben-Ghedalia, D.; Miront, J. Effect of lactic acid bacteria inoculants on in vitro digestibility of wheat and corn silages. Journal Dairy Science, v. 90, pp. 4753–4762, 2007. DOI:10.3168/jds.2007-0176

[38] Fuller, R. Probiotics in man and animal. Journal Applied Bacteriology, v. 66, pp. 365–378, 1989.

[39] Aksu, T.; Baytok, E.; Bolat, D. Effects of a bacterial silage inoculant on corn silage fermentation and nutrient digestibility. Small Ruminant Research, v. 55, pp. 249–252, 2004. DOI: 10.1016/j.smallrumres.2003.12.012

[40] Gollop, N.; Zakin, V.; Weinberg, Z.G. Antibacterial activity of lactic acid bacteria included in inoculants for silage and in silages treated with these inoculants. Journal of Applied Microbiology, v. 98, 662–666, 2005. DOI:10.1111/j.1365-2672.2004.02504.x

[41] Weinberg, Z.G.; Muck, R.E.; Weimer, P.J. The survival of silage inoculant lactic acid bacteria in rumen fluid. Journal of Applied Microbiology, v. 94, 1066–1071, 2003.

[42] Amado, I.R.; Fuciños, C.; Fajardo, P.; Guerra, N.P.; Pastrana, L. Evaluation of two bacteriocin-producing probiotic lactic acid bacteria as inoculants for controlling Listeria monocytogenes in grass and maize silages. Animal Feed Science and Technology, v. 175, pp. 137–149, 2012. DOI: http://dx.doi.org/10.1016/j.anifeedsci.2012.05.006

[43] Nsereko, V.L.; Smiley, B.K.; Rutheford, W.M.; Spielbauer, A.; Forrester, K.J.; Hettinger, G.H.; Harman, E.K.; Harman, B.R. Influence of inoculating forage with lactic acid bacterial strains that produce ferulate esterase on ensilage and ruminal degradation of fiber. Animal Feed Science and Technology, v. 145, pp. 122–135, 2008. DOI: http://dx.doi.org/10.1016/j.anifeedsci.2007.06.039

[44] Fugita, C.A.; Prado, I.N.; jobim, C.C.; Zawadzki, F.;Valero, M.V.; Pires, M.C.O.; Prado, R.M.; Françozo, M.C. Corn silage with and without enzyme bacteria inoculants on

performance, carcass characteristics and meat quality in feedlot finished crossbred bulls. Revista Brasileira de Zootecnia, v.41, n.1, pp. 154–163, 2012.

[45] Zopollatto, M.; Daniel, J.L.P.; Nussio, L.G. Microbial additives in silages in Brazil: review of aspects of silage and animal performance. Revista Brasileira de Zootecnia, v. 38, pp. 170–189, 2009. ISSN on-line: 1806-9290

[46] Muck, R.E. Silage additives and management issues. Proceedings of Idaho Alfafa Forage Conference, Best Western Burley Inn, Burley, Idaho, USA, pp. 49–55.

[47] Faber, D.A.; Linn, J.G.; Otterby, D.E. Effect of a bacterial inoculant on the fermentation of high moisture shelled and ear corn. Journal Dairy Science, v. 72, pp. 1234–1242.

[48] Neumann, M.; Oliboni, R.; Oliveira, R.M.; Faria, M.V.; Ueno, R.K.; Reinerh, L.L.; Durman, T. Chemical additives used in silage. Pesquisa aplicada & Agrotecnologia, v. 3, n. 2, mai-ago, 2010. ISSN 1984-7548

[49] Rosa, B.; Fadel, R. Use of anhydrous ammonia and urea to improve the nutritional value of conserved forage. Anais do Simpósio sobre Produção e Utilização de Forragens Conservadas, pp. 41–63, 2001.

[50] Schmidt, P.; Mari, L.J.; Nussio, L.G.; Pedroso, A.D.F; Paziani, S.D.F.; Wechsler, F.S. Chemical and biological additives in sugarcane silage. 1. Chemical composition of silages, intake, digestibility and feeding behavior. Revista Brasileira de Zootecnia, v. 36, n. 5, pp. 1666–1675, 2007.

[51] Williams, P.E.V.; Innes, G.M.; Brewer, A. Ammonia treatment of straw via hydrolysis of urea. Effects of dry matter and urea concentration on the rate of hydrolysis of urea. Animal Feed Science Technology, v. 11, n. 2, pp. 115–124, 1984. 0377-8401

[52] Fadel, R.; Rosa, B.; Oliveira, I.P.; Oliveira, J.D.S. Evaluation of different proportions of water and urea on the chemical composition of rice straw. Ciência Animal Brasileira, v. 4, n. 2, pp. 101–107, 2003.

[53] Tarkov, H., Feist, W.C. A mechanism for improving the digestibility of lignocellulosic material with dilute alkali and liquid ammonia. Advanced Chemistry Series, v. 26, n. 1, pp. 13–21, 1969.

[54] Chen, L.; Guo, G.; Yuan, X.; Shimojo, M.; Yu, C.; Shao, T. Effect of applying molasses and propionic acid on fermentation quality and aerobic stability of total mixed ration silage prepared with whole-plant corn in Tibet. Asian Australian Journal Animal Science, v. 27, n. 3, pp. 349–356, 2014 DOI: http://dx.doi.org/10.5713/ajas.2013.13378

[55] Selwet, M. Effect of propionic and formic acid mixtures on the fermentation, fungi development and aerobic stability of maize silage. Polish Journal of Agronomy, v.1, pp. 37–42, 2009.

[56] Kung Jr., L.; Sheperd, A.C.; Smagala, A.M.; Endres, K.M.; Besset, C.A.; Ranjit, N.K.; Glancey, J.L. The effect of preservatives based on propionic acid on the fermentation

and aerobic stability of corn silage and a total mixed ration. Journal Dairy Science, v. 81, pp. 1322–1330, 1998.

[57] Amer, S.; Hassanat, F.; Berthiaum, P.; Mustafa, A.F. Effects of water soluble carbohydrate conten on ensiling characteristics, chemical composition and in vitro gas production of forage millet and forage sorghum silages. Animal Feed Science and Technology, v.177, pp. 23–29, 2012. http://dx.doi.org/10.1016/j.anifeedsci.2012.07.024.

[58] Guimarães Jr., R.; Gonçalves, L.C.; Rodrigues, J.A.S.; Borges, A.L.C.C.; Noberto, M.R.; Saliba, E.O.S.; Pires, D.A.A.; Jayme, D.G.; Castro, G.H.F.; Fibrous fraction of original materials and silages of three millet genotypes (*Penissetum glaucum* (L). R. Br.). on different fermentation periods. Revista Brasileira de Milho e Sorgo, v.4, n.2, pp. 243–250, 2005. http://dx.doi.org/10.18512/1980-6477/rbms.v4n2p243–250.

[59] Costa, K.A.P.; Assis, R.L.; Guimarães, K.C.; Severiano, E.C.; Assis Neto, j.M.; Crunivel, W.S.; Garcia, J.F.; Santos, N.F.; Silage quality of brachiaria brizantha cultivars ensiled with different levels of millet meal. Arquivo Brasileiro de Medicina Veterinária e Zootecnia, v.63, n.1, pp. 188–195, 2011.

Chapter 7

Maximizing Fiber Utilization of Silage in Ruminants

Basim Refat and Peiqiang Yu

Additional information is available at the end of the chapter

http://dx.doi.org/10.5772/64471

Abstract

This chapter highlights the importance of fiber digestibility and utilization in ruminants and to summarize the main factors that influence fiber digestibility in silages. Forage provides at least half of the diet of lactating cattle and greatly affects energy and carbohydrate intake. It is important to maximize the intake of digestible carbohydrate from forages, because energy requirements for maintenance and milk production often exceed the amount of energy high-producing cows can consume, particularly in early lactation. There are many approaches used for enhancing fiber utilization in silage and subsequent maximizing energy intake and productivity of dairy cattle. Out of these approaches are: selecting appropriate forages with high fiber digestibility, applying the appropriate agronomic practices such as harvesting at the proper stage of maturity, fertilization, and cutting height at harvest, along with using of esterase-producing inoculants or fibrolytic enzymes have been proposed as approaches to improving the productivity of dairy cattle.

Keywords: feed additive, fiber utilization, nutrient availability, ruminants

1. Introduction

The global livestock industry faces an extensive challenge since a presumed dichotomy exists between the increasing requirements for animal feeding conferred by population growth and consumer concerns regarding the sustainability of livestock production [1]. Meanwhile, the cost of feed grains for livestock has increased substantially in recent years [2]. Thus, there is an increasing interest in using silages as a main source of forages in ruminant's diets, with high nutritive value as an alternative feed source. In high-producing dairy cattle, it is important to maximize digestible carbohydrate intake or increase neutral detergent fiber digestibility (NDFD) from silage because the energy needed for maintenance and milk production often exceeds the amount of energy high-producing cows can consume, particularly in early lactation [3]. One of the main factors that affect silage utilization is the proportion of its potentially digestible fiber fraction, where silage having less than 60% of total fiber content is available for digestion by the ruminant animal [4]. The first section of this chapter will discuss the most important aspects of silage fiber digestibility. The chapter starts by the importance of fiber digestibility, before considering the method used for evaluating fiber digestibility. This is followed by fiber digestion and utilization in ruminants. The chapter ends with sections on the factors that effect on fiber digestibility in silages.

2. Importance of fiber digestibility

Silages are considered the most cost-effective feed resource in ruminant nutrition. Grass and small-grain cereal silages are the main sources of dietary energy, while leguminous silages are considered important sources of protein for ruminant livestock [5]. The quality of silage is an important determining factor in dairy cow performance as the forage accounts for a large proportion of the diet about reaching from 35% up to 100% of dry matter (DM) [6]. For high-producing dairy cows, high-quality silages with lower fiber and higher fermentable concentrates are usually used to meet energy requirements. Nevertheless, inadequate dietary fiber reduces chewing activity, insalivation and rumen pH, and can cause rumen acidosis and laminitis [7]. These can depress fibrolytic microbes and milk production by increasing maintenance demands [8, 9]. National Research Council (NRC) stated that dairy rations should have a minimum of 25% neutral detergent Fiber (NDF), 18.7% of which must come from forage for adequate rumen health. Although rumen fermentation and function can cause negative impacts on dairy cattle fed rations deficient in fiber, excessive level fiber of over 44% may also have negative effects on intake and digestibility [9].

The National Research Council (NRC) recommendations regarding the total NDF and forage NDF contents of dairy rations are presented in **Table 1** [9]. In general, the minimum NDF contents that are recommended for dairy ration will depend on the dietary contents of NFC, a physical effectiveness of fiber, and the source of the fiber. It is well established that the fiber from forage sources could induce the salivation and cud-chewing activity than nonforage fiber sources. Consequently, the major factor for evaluating the efficiency of dietary NDF capability is NDF content in forages. It has become very important to prevent acute and subacute rumen acidosis and maintain milk fat level, evaluating the physical effective NDF (peNDF) in diets due to the importance of peNDF in maintaining the rumen pH and fiber digestion. It is well established that the amount of peNDF in the diet is dependent on the chop length of forages, dietary NDF, and forage to concentrate ration content [10]. It has been reported that peNDF intake can stimulate the chewing activity and can minimize the incidence of ruminal acidosis [11]. Many studies have examined the effects of peNDF on lactation performance [12–19]. The peNDF of feed could be calculated from the NDF content multiplied by a physical effectiveness factor (pef). The pef ranges between 0 (not effective at stimulating chewing) and 1 (100% effective at stimulating chewing). Numerous feed models such as Cornell Net Carbohydrate and Protein System (CNCPS) presently use peNDF as an important input for the model to

http://dx.doi.org/10.5772/64471

predict lactational performance. The forage and total mixed ration (TMR) particle size distribution recommendation using Penn state particle separator as reported by Heinrichs and Kononoff is presented in **Table 2** [13].

Minimum NDF from forage	*NDF from forage (% of total NDF)*	*Minimum NDF in diet*
19	75	25
18	66	27
17	58	29
16	51	31
15[a]	45	33

[a] Not recommended because of depression of milk fat test.

Table 1. Recommended minimum NDF concentration based on the proportion of NDF coming from forage sources [9].

Sieve size	**Type**		
	Corn silage	**Haylage**	**TMR**
>19.0 mm	5 ± 3	15 ± 5	5 ± 3
19.0–8.0 mm	55 ± 10	60 ± 15	40 ± 10
8.0–1.18 mm	40 ± 10	30 ± 10	40 ± 10
<1.18 mm	<5	<5	<20

Table 2. Forage and TMR particle size distribution using Penn state particle separator as reported by Heinrichs and Kononoff [13].

3. Evaluating of fiber digestibility in ruminants

Understanding the mechanism of fiber digestion is very important to accurately estimate the digestible energy of fiber and to improve animal performance. Fiber is digested primarily in the rumen as the result of the dynamic operation that is affected by the chemical nature of the fiber and by the passage and digestion rate of fiber within the digestive tract of the animal. The potentially digestible NDF (pdNDF) and the digestion rate (kd) vary greatly between and within different silage types [14, 17]. The passage rate of fiber (kp) is in the first place influenced by the animal, where the digestion of fiber increases along with increased retention time of feed in the rumen [15, 18]. Several models have been developed to describe the process of digestion in the rumen; some models are simple or complex. Most of these models have been developed by fractional schemes to correlate the disappearance or gas production curves with rumen digestibility of feed components, which assume that the feed component includes at least two portions: a potentially degradable fraction and an undegradable fraction. The potentially degradable portion will be degraded at a fractional rate (per hour), after a discrete lag time (h). The undegradable fraction is calculated from the longer time of incubation as proposed by Waldo et al. [19] (**Figure 1**). By using this model, Allen and Mertens [21] educed mathematical equations to define fiber digestibility and rumen fill. For fiber digestibility, the following equations were educed:

$$D=\left(pdNDF\left(dFINTAKE/dt\right)\right)/(kd+kp) \tag{1}$$

$$I=\left(fi\left(dFINTAKE\,/\,dt\right)\right)/\left(kp\right) \tag{2}$$

Finally, the rumen fill would be estimated as the sum of the digestible (D) and indigestible (I) fiber pools in the rumen

$$Fill=D+I \tag{3}$$

Eq. (1) shows that digestibility is directly related to (pdNDF) and (kd), and inversely proportional (kd + kp; the rate of total fiber digestibility). Thus, as the ruminal retention time increases (1/kp), the extent of ruminal digestibility increases [22]. The fiber weight in the rumen is dependent on fiber intake per unit of time (dFINTAKE/dt), and parts that are digestible (fd),

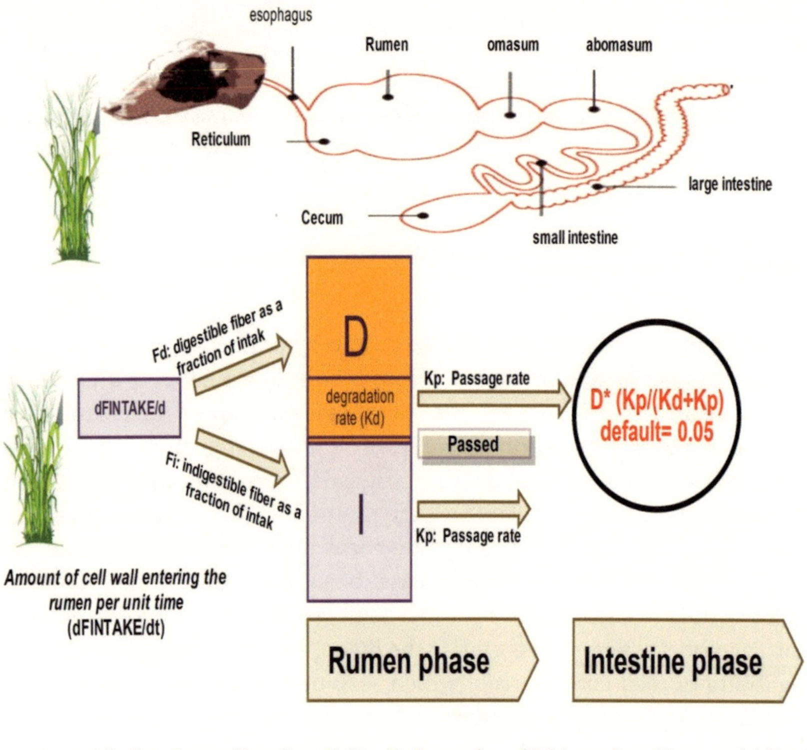

Figure 1. Schematic model of total-tract fiber digestibility. Redrawn from Waldo et al. and Jung and Allen [19, 20].

and indigestible (fi), as well as digestion rates (kd) and passage (kp). Jung and Allen ranked the factors that influence ruminal fill, and the most important element was the fiber content, followed by kp, the fraction that is indigestible, and the lowest factor was the kd [20]. The digestion kinetics of fiber can be measured in vivo using rumen evacuation technique, where cannulated animals are used for measuring the digestible and indigestible fiber pools that flow from the rumen [23]. In spite of the high precision for rumen evacuation technique to estimate rumen digestion kinetics, this technique is unwieldy for routine forage analysis. It has been proved that the use of other biological methods, that is, in vitro or in situ techniques, could give better characterization to degradation kinetics of fibrous fraction of forages. Over the last 50 years, the in vitro system has not been widely used in farm to implement analysis on forages because of its difficulty to perform in farm. This situation has changed in recent years with the use of a shorter digestion time (30 or 48 h) along with the enhancements that occurred in spectral analysis using near-infrared spectroscopies, where the laboratories were facilitated to assess the digestion of forages without the need to obtain rumen fluid. Some mathematical equations have been developed, which can use single time points like 24 or 30 h in vitro NDFD along with fixed lag time and lignin in the forages to calculate the kd rates [24].

In recent times, the feeding studies have found the indigestible neutral detergent fiber (iNDF) after longer incubation time (240 h in vitro or 288 h in situ) was highly correlated with dry matter intake (DMI) and would be used to predict pdNDF [25]. Furthermore, there were sufficient data being created by commercial laboratories. Thus, the iNDF was applied as a new approach rather than using lignin × 2.4 to calculate pdNDF (CB3) and indigestible NDF (CC) using the updated CNCPS 6.5 [25]. It has been found that the model, which could accurately predict NDF digestibility, should partition NDF into iNDF and pdNDF, fractionate feed particles by their retention and passage in the rumen, using a predicted kd by an in vitro system [26]. Based on this approach, Combs developed a new method for predicting fiber digestibility; he used shorter incubation time (24, 30, and 48 h) along with iNDF (240 h) to predict kd (kdCB3) of pdNDF [27]. The CB3 kd rates derived from in vitro analysis were entered in the updated CNCPS model to calculate the ruminal fiber digestibility according to this equation; rumen degradability for pdNDF = CB3 × (kdCB3/(kdCB3 + kp). Finally, they calculated the in vitro total-tract NDFD (ivttNDFD) assuming that the intestinal digestibility of available NDF (CB3) amount escaping rumen digestion was 5%. Lopes et al. have found that in vivo total-tract NDF digestibility was highly correlated with the ivttNDFD. The regression equation to describe the relationship was described as follows: in vivo total-tract NFDF (%) = −3.62 + 1.11 × ivttNDFD (%) with $R^2 = 0.70$, RMS = 4.27, P-value < 0.01; $n = 21$ diets. The differences between two methods (ivttNDFD and in vivo total-tract NDFD) were not significant, and mean values varied by only 1% unit, showing promise for this approach [28].

The use of high-resolution spectroscopic techniques (e.g., high-field nuclear magnetic resonance, mid-infrared, Raman spectroscopy, and pyrolysis mass spectrometry) is finding increased usage in forage assessment. These advanced technologies would provide more broad information about a primary nature [29]. A spectroscopic method such as Fourier transform infrared (FT/IR) spectroscopy has been developed as rapid, direct, nondestructive and noninvasive bioanalytical technique [29–37]. Thereby, this technique paves the way to better understand the quantity, composition, structure, and distribution of chemical constituents and functional groups in a tissue (feed and ingredients) [38–42]. Intrinsic chemical structures were found to effect on nutritive value, degradation characteristics, utilization, and availability of feed [43, 44]. Many studies have reported that AT/IR would accurately predict rumen degradability of DM, NDF, concentrations of lignin, ferulic, and coumaric acids in forage samples [45–47].

4. Fiber digestion and utilization in ruminants

4.1. Plant cell-wall carbohydrates

The forages are diverse in its characteristics, and this uniformity results in variations in quality as an animal feed. Plant cell-wall carbohydrates are the most important components in forages that influence silage quality. There is higher complexity in the utilization of silages due to diversity among forage plants, diversity in the ruminal microorganisms, and interaction between the forage plant cell-wall carbohydrates and microorganisms [48]. Ruminants can digest and degrade plant cell-wall polysaccharides. The plant cell-wall chemistry and anatomical structure will determine the digestion characteristics of cell types [49]. The fiber fraction for the main silages is presented in **Table 3**.

Forage	% DM	ADF	NDF	Hemicellulose	Lignin
Legume silage	37	39	47	8.9	7.7
	30–43	33–44	40–55	4.1–13.6	5.3–10.0
MM legume silage[a]	35	39	52	13.4	6.8
	27–42	35–42	45–59	7.8–18.9	5.4–8.3
MM grass silage	36	39	56	17	6.9
	28–45	35–44	50–63	22	4.7–9.0
Grass silage	31	41	62	21	6.4
	21–41	37–44	55–68	15–27	4.9–7.8
Corn silage	33	26	45	19	2.8
	25–40	22–30	38–51	15–23	2.2–3.5
Winter cereals	29	31	52	21	4.3
	35	39	59	20	6.3

[a]MM legume refers to mixed mainly legume forage; MM grass refers to mixed mainly grass forage.

Table 3. Fiber fraction for NDF concentrations based on the proportion of NDF derived from forage sources

The main groups of plant cell-wall carbohydrates are hemicelluloses and cellulose. Cellulose is a water-insoluble β-glucan composed of a linear molecule of D-anhydroglucopyranose residues linked by a β-(1→4) bond. In contrary to cellulose, hemicellulose has various groups of polymers that are characterized with the heterogeneous composition. Xylan is the main component of hemicellulose and compromises about 30–35% of the cell-wall material of annual plants. The main chain of xylan is composed of 1,4-β-linked D-xylopyranose units [50, 51].

The collaborative activity of the cellulolytic and noncellulolytic microorganisms in the rumen is critical in fiber digestion [52]. Rumen cell-wall degradation initiated by the attachment of rumen microbes to fiber and the bacterial species specialized to start this attachment/colonization process are the cellulolytic species *Ruminococcus albus*, *R. flavefaciens*, and *Fibrobacter succinogenes*. Rumen fungi and protozoa also colonize and degrade plant fragments to differing degrees [48]. The fermentation of structural carbohydrates by cellulolytic consortium results in the progressive process where volatile fatty acids (VFAs) are liberated at a lower rate than starch fermentation. The fermentation of structural carbohydrates is associated with an increase in the proportion of acetic and butyric acid [53]. Following absorption, the large proportion of acetate is not changed by hepatic metabolism and may be augmented by endogenous acetate production in the liver. The posthepatic supply of acetate to peripheral tissues constitutes a major part of the total energy available to the animal and may be either oxidized to produce adenosine triphosphate (ATP) or used as a substrate in the production of long-chain fatty acids [54]. While ruminally derived butyrate is quantitatively metabolized to b-OH-butyrate during absorption through the rumen epithelium, in posthepatic tissues it has a similar metabolic fate to that of acetate [54].

4.2. Lignin and phenolic acids

Lignin is an indigestible polymer in plants that plays an important role in the structural integrity of plant tissue. Although lignin comprises little of the total structural carbohydrate system in plants, it has been recognized to exert the negative effect on cell-wall polysaccharide digestibility by coating the plant cell-wall polysaccharides from enzymatic hydrolysis [55]. Lignin arises from an enzyme-initiated dehydrogenative polymerization of three originators: p-coumaryl alcohols, coniferyl, and sinapyl. The phenylpropanoid metabolism and shikimic acid pathway lead to the synthesis of lignin intermediates like p-coumaric acid, ferulic acid, and diferulic acid [56], which are converted into coniferyl, sinapyl, and p-coumaryl alcohols and ultimately to guaiacyl, syringyl, or p-hydroxyphenyl lignin, respectively [55].

With the maturation of forage cell walls, the guaiacyl-type lignin changes to lignin-rich syringyl units, and the digestibility of mature cell walls decreased. Taboada et al. found that guaiacyl and syringyl have negative correlation with organic matter or dry matter digestibility in ruminants fed on silages. They concluded that guaiacyl and syringyl could be used as predictors of digestibility than total lignin content in silage [57].

The brown midrib (BMR) mutation in annual C4 grasses such as corn and sorghum results in both a reduction in lignin concentration and a shift in lignin composition to a more guaiacyl-rich polymer [20]. Jung and Deetz have suggested that the improved digestibility of cell walls in BMR mutants is a result of both the reduced lignin concentration and the reduction in syringyl lignin content [58].

Cross-linking of lignin to cell-wall polysaccharides has been reported as additional mechanisms limiting fiber digestibility [20]. In grasses, ferulate and p-coumarate molecules are esterified to arabinoxylans, and some of p-coumarates are the ester or covalent linked to lignin [59]. As forages mature and lignin concentrations increase, ferulates that were esterified to arabinoxylan become etherified to lignin via cross-links between lignin and the cell-wall polysaccharides [60]. The degree of lignin/arabinoxylan cross-linking by ferulates negatively influences cell-wall digestibility to the polysaccharides, which prevents physical access by hydrolytic microbial enzymes to polysaccharides [49]. Model studies utilizing isolated cellulose and xylans, and forage NDF to which phenolic acids have been synthetically esterified, obviously demonstrated that the presence of these phenolic esters negatively effects on cell-wall degradability [61]. However, the reduction in digestibility caused by esterified ferulic acid only limits the degradation rate of polysaccharide, rather than extent, because fungi and ruminal bacteria possess phenolic acid esterases to ultimately remove these impediments to cell-wall digestion [62].

5. Enhancing fiber digestibility and utilization of silage

Ruminal digestibility of forage neutral detergent fiber can range from less than 25% to over 75% for different forage types [9]. Most research with brown midrib mutant corn silage found that lactating dairy cows will consume more DM and produce more milk when fed corn silages that have greater NDFD [63–65]. Oba and Allen found a relationship between NDFD and animal performance and they reported that a 1-unit increase in forage NDFD after 30 h of in vitro incubation was associated with increases of 0.17 kg d^{-1} of dry matter intake, 0.23 kg d^{-1} of milk yield, and 0.25 kg d^{-1} of 4.0% fat-corrected milk [66]. Using high-quality silage in dairy cattle rations could reduce physical rumen fill, allow cattle to consume more feed, and produce more milk [63]. There are many factors that would influence the quality of silage. Such factors include silage species, silage varieties, stage of harvest, cutting height, growing conditions, silage additives, and enzymes.

5.1. Silages species

The most practical approach for increasing NDFD is based on increasing the amount of pdNDF in forages. Grass silages often have a greater proportion of pdNDF to indigestible NDF (iNDF) and higher in NDFD than legume silages, but the rate of digestion of legume pdNDF is frequently faster and could increase the total amount of NDF digested in vivo [63, 64]. The chemical and structural features have been identified, which may reduce the fiber digestion. Of these, lignin is the most notably reported [67]. Lignin is supposed to constrain ruminal fiber digestion, which acts as a physical barrier. The involvement of cross-linking of lignin to polysaccharides by ferulate linkages as an additional factor that inhibits the digestion of grass fibers has been identified [20]. However, a similar lignin cross-linking to fiber polysaccharides in legumes has not yet determined. There is an important role for plant anatomy on fiber digestibility [68]. The vascular tissue, sclerenchyma, and stem epidermis are degraded at a slower rate in rumen where they contain a higher amount of indigestible or highly lignified components. Leaf blades C4 grasses are typically less digestible than those in C3 grasses due to the existence of mesophyll cells. In C3 species, stem tissue cell such as parenchyma bundle sheath, mesophyll, phloem, and epidermal cells are totally degraded, but these tissues are partially or slowly degraded in C4 species. In an earlier study by Akin and Burdick, they found that C4 grasses are less digestible than C3 species due to the existence of vascular tissue and parenchyma bundle sheath cells in larger amounts than in C3 grasses [69].

The total-tract digestibility of whole-crop cereals silage, legumes, and maize silage is often lower than for grass silage. However, the lower digestibility is mostly alleviated by higher feed intake such that energy intake is maintained [70]. Many studies have shown that the partial replacement of grass silage with whole-crop cereals may not have a negative impact on milk production in cows [71]. However, the effects of barley silage on DMI have been inconsistent, which are probably attributable to differences in the quality of the forages between studies. For example, Ahvenjärvi et al. noted a reduction in fiber digestibility when grass silage was replaced with whole-crop barley silage. This reduction in NDFD was related to a lesser pdNDF concentration in the rumen and higher iNDF pool size of barley silage compared with that of grass silage [70].

Whole-crop cereals species also varies in their quality and digestibility, for example, barley and oat silages when harvested at the same maturity stage (milk to soft dough stage) have found to enhance the feed intake and average daily gain in heifers when compared with triticale silage [72]. Furthermore, dairy cows that fed on barley silage have had higher intake than cows fed on oat silage when harvested at the maturity stage (early to a mid-dough stage of maturity). Such difference in feed intake is a consequence of variation in chemical composition and ear:stalk ratio of whole-crop cereals. Barley has more starch than oats and triticale because of the higher ear:stalk ratio in barley. Since most fibers exist in plant stalk, barley contains a lower fiber than oats and triticale when they are harvested at the stage of maturity. The higher starch resulted in a lower fiber content in barley silage, and hence barley can enhance the OM digestion when compared with oats and triticale silages when fed to dairy cows [72].

5.2. Selecting varieties with enhanced NDFD

Another potential method to increase pdNDF is by the use of genetic mutations in forage crops that reduce iNDF and increase the pdNDF fraction of the plant. The brown midrib mutation mutants were discovered for the first time at the University of Minnesota in 1924; the BMR genes have been found in sorghum, Sudan grass, millet, and corn. The BMR corn forage has about 25% less lignin and lower cross-linkages with lignin. Corn silage with the brown midrib mutation has a higher NDFD (34% less lignin and had 19% higher IVNDFD than conventional corn silage) [73–75]. Several studies confirmed the positive effect of feeding BMR corn on DMI and productivity of dairy cattle [76, 77], but responses have not been consistent in all experiments [78]. Ivan et al. compared corn silage with low and high cell-wall content on milk production, and reported that the hybrid with high cell-wall content had greater IVNDFD, increasing DMI and milk yield [79]. Data collected from a Journal of Dairy Science (number of treatments $n = 22$; **Table 4**) between the year 1999 and 2010 showed a non-significant correlation between IVNDFD in BMR corn silage and milk yield or DMI ($P > 0.05$, **Figures 2** and **3**). Inconsistent results between experiments may be attributed to various factors such as including cows at a different stage of lactation and duration of experimentation or the lack of effect of forages with enhanced NDFD on DMI [82]. Recently, Ferraretto and Shaver performed a meta-analysis to study the effect of corn silage hybrids with different stalk characteristics (conventional, dual-purpose, isogenic, or low-normal fiber digestibility, brown midrib, hybrids with greater NDF but lower lignin contents or high in vitro NDF digestibility, and leafy corn silages) on lactation performance [85]. They found that for every 1-unit increase in ivNDFD the DMI can increase by 0.09 kg/d, although this correlation was not significant (DMI = 0.09ivNDFD + 19.531; $R^2 = 0.72$, $p = 0.40$); additionally, they found that for every 1-unit increase in ivNDFD the milk yield would increase by 0.14 kg/d (milk yield = 0.14 ivNDFD + 31; $R^2 = 0.87$, $P = 0.06$). It has been reported that the total-tract NDFD response to feeding bm3 corn silage is influenced by the DMI response due to enhanced ivNDFD as reported by Oba and Allen [64]. On the other hand, corn silage type, that is, bm3 versus near-isogenic or conventional corn silage hybrids by dietary forage NDF [82], starch [65], and CP [76] concentration, or supplemental corn grain endosperm type [80] interactions were undetected.

Publication	*Treatments (n = 22)*[a]
Ballard et al. [81]	Mycogen corn silage
	Cargill (brown midrib corn silage)
Castro et al. [82]	Normal corn silage
	Brown midrib corn silage
Ebling and Kung, Jr. [83]	Conventional corn silage
	Brown midrib corn silage
Gehman et al. [78]	Dual-purpose corn silage
	Brown midrib corn silage
Ivan et al. [79]	Corn silage with lower cell-wall content
	Corn silage with higher cell-wall content
Oba and Allen [65]	Control corn silage
	Brown midrib corn silage
Oba and Allen [66]	Control corn silage
	Brown midrib corn silage
Taylor and Allen [80]	Control corn silage
	Brown midrib corn silage
Thomas et al. [84]	Dual-purpose corn hybrid
	Leafy corn silage hybrid
Weiss and Wyatt [76]	Dual-purpose corn silage
	High fiber corn silage
Weiss and Wyatt [76]	Dual-purpose corn silage
	Brown midrib corn silage

[a]Correlation analysis between the two variables was performed using the CORR procedure of SAS with the Pearson correlation method, because the variable data are normally distributed. Average of milk yield (38.2 ± 4.360), average of IVNDFD (50.39 ± 9.162).

Table 4. Effects of silage varieties with enhanced 30-h IVNDFD on milk yield. Data have been taken from a number of publications in Journal of Dairy Science (JDS from 1999 to 2010).

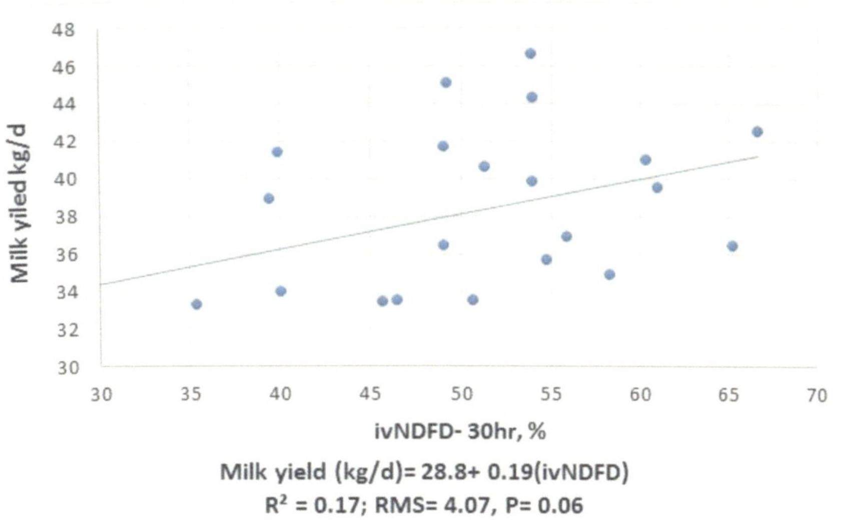

Figure 2. Relationship between in vitro NDFD (30 h) and milk yield with the prediction equation.

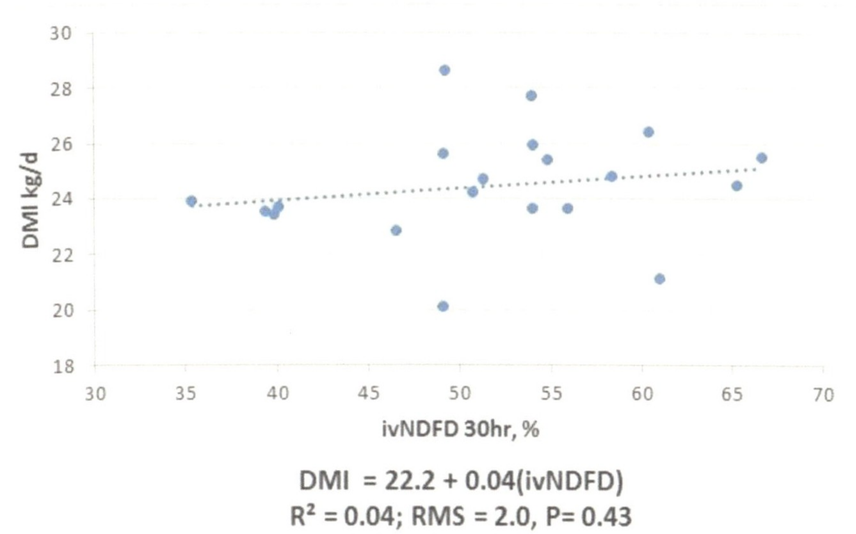

Figure 3. Relationship between in vitro NDFD (30 h) and DMI with the prediction equation.

5.3. Agronomic practices to enhance fiber digestibility

Fiber digestibility is largely dependent on plant maturity. The effect of harvest maturity of whole-crop annual forages is more variable concerning fiber content. Rosser et al. reported a reduction in NDF content by advancing the maturity of barley and oat forage from head elongation to fully ripe, with a reduction in NDF content from 13.8 to 9.6% [86, 87]. By contrast, the NDF concentration of whole-crop barley was not changed during the milk and soft dough stages, but it increased somewhat between the soft and hard dough stages while this change was not observed in whole-crop oat forage [88]. Bolsen and Berger reported a reduction in total-tract DM digestibility of barley silage at milk stage, compared to advanced, mature stage due to the increasing grain content [89]. By contrast, Rustas et al. found no changes in DM or NDF digestibilities for wheat forage ensiled at milk and dough stages. However, the response regarding NDF digestibility varied for barley forage that was ensiled at milk and dough stages depending on location [89].

With advancing the maturity of grasses silage, their digestibility dramatically drops because the tensile strength of stems increases to support the weight of the plant, besides the leaf-to-stem ratio declines [15, 18]. In grass silage, organic matter digestibility dropped from 79% in early growth to 73% in late growth, and NDFD decreased from 73% in early growth to 66% when the plant maturity reached late growth stage. In legumes, NDFD is less than the grasses or small grains during the early vegetative stage of growth but drops slower with advancing maturity.

In corn silage, the stage of maturity has an impact on fiber fraction. The fibrous content has been observed to decline with increasing maturity in whole-corn plants, but no significant change in lignin concentration from early dent to black layer [90]. Coors et al. suggested the observed drop in fiber concentration with increasing maturity to the dilution effect with increasing percentage of grain as the corn plant matures [91]. Fiber concentration of corn stover increases as maturity increases [92, 93].

Increasing the height of cutting, which results in leaving a larger proportion of less digestible stalk in the field, may increase the feeding value of silage for lactating dairy cows. It has been reported that corn silage digestibility was enhanced at cutting heights of 45–50 cm. but this at the expense of DM yield [94, 95]. Kruczyńska et al. reported a reduction in hemicellulose, cellulose, and lignin and greater effective degradability of silage that was cut at 50 versus 10 cm [96]. Neylon and Kung examined the effects of corn plant-cutting height and maturity on silage nutrient value. Plants were cut at 12.7 and 45.7 cm as well as harvested between one-third and two-third milk line and then again at black layer [97, 98]. As anticipated, NDF tended to be less in silages that were cut higher, and ADF content decreased significantly. At later maturity, the lignin contents were not influenced by increasing cutting height. The cutting height only influenced in vitro NDF digestibility, with the higher cut being more digestible. By increasing the cutting height of corn silage, the nutritive value was increased by decreasing NDF, ADF, and acid detergent lignin concentration and increasing the starch concentration. They also found that as corn plants were cut higher, there was a tendency for increased milk production and increased feed efficiency in dairy cows. Kung et al. also observed a decrease in fiber fraction concentrations, as well as an increase in starch, and crude protein concentrations as cutting height, was increased [97, 98]. These observations are all logical, because when cutting height is increased, more lignified and less digestible stems are left in the field while increasing the concentration of more digestible leaves and kernels.

It is well established that the nitrogen fertilization can increase the protein content and forage yield and decrease the fiber content. Campos et al. reported a reduction in hemicellulose content and arabinose proportion of the fiber fraction in Milenio grass by N fertilization. They also found that the fertilization increased fiber digestibility due to increase in (arabinose + glucose):xylose ratio [99].

Environmental temperature has a significant impact on forage digestibility. The forages grown under higher environmental temperature had the higher amount of lignin [100]. Altering the time of seeding can shift the stage of maturity when plants are exposed to greater ambient temperature, moisture availability, and photoperiod intensity. Chow et al. found that the exposure of forages to a lower environmental temperature during heading stage increased IVNDFD [101].

5.4. Silage inoculants

Silage inoculants can be added to the freshly harvested forages to obtain good-quality silage. The first studies on adding inoculants for improving the quality of silage used the inoculants that contain homolactic bacteria (LAB) such as *Lactobacillus plantarum*, which quicken the drop in silage pH. Nevertheless, this rapid drop in pH inhibits the growth of yeasts, spoilage bacteria, and fungi, as well as plant cell breathing, maintaining the sugars in the silage without decomposition [102]. If this happens, the yeast consumes the lactic acid for its growth causing an augment in silage pH. At this stage, each of yeast and mold can quickly take advantage of sugars for their growth, and reduce the density of nutrients in silage. Due to the occurrence of losses in silage-nutrient density, the studies on developing the inoculant production came up with the second-generation silage inoculants that were generated from *Propionibacteria* spp. and *L. buchneri* [102, 103]. Overall, studies have shown that *buchneri L.* inoculants are more effective in improving aerobic stability of silage than *Propionibacteria* inoculant. *Lactobacillus buchneri* is one of heterolactic bacteria, which is able to ferment lactic acid to acetic acid; the acetic acid in turn has an inhibitory effect on the growth of yeast and subsequently prolong the silage shelf life and reduce deterioration of silage nutrients [104]. It was proposed that *L. buchneri* inoculation would reduce feed intake in ruminant livestock as a result of acetic acid production. However, no effect of inoculant on feed intake has been reported when *L. buchneri*-treated silage has been fed [105–109].

The first and second generation of inoculants focused only on improving the silage stability without addressing improving the nutrient availability by animals. The main reason for the limited effect in the first and second generation was the inoculants did not produce enzymes that digest the plant cell walls. Thus, the third-generation silage was introduced more recently, through feeding silage inoculated with lactic acid bacteria with ferulic acid esterases activity. Previous studies by Yu et al. have shown that *Aspergillus* ferulic acid esterase and *Trichoderma xylanase* act synergistically to release ferulic acid from feruloyl-polysaccharides in complex plant cell walls [110, 111]. This activity opens the rest of the polysaccharides for more hydrolytic attack and facilitates the accessibility of the main polysaccharide chain to cellulase, thereby increasing the release of reducing sugars [110, 111]. Nsereko et al. performed a screening study on 1000 esterase-producing *Lactobacillus* bacteria and found that half of this number could be able to produce ferulic acid esterase, and run more detailed studies on eight of the bacteria. When compared to untreated perennial ryegrass, all inoculated samples had 9–11% greater

NDFD. Moreover, they found that the inoculation of four corn silage hybrids with a combination of *L. buchneri* and *L. paracasei tolerans* enhanced NDFD by 7% [112, 113]. Several studies have confirmed that esterase enzymes can complement the effects of cellulose and hemicellulase enzymes on plant cell walls, thereby increasing DM or fiber digestibility [114]. Conversely, some studies have reported no effect from adding ferulic acid esterase-producing inoculant on fiber digestibility of silage [115]. Kang et al. reported an enhancement in fiber digestibility when corn hybrids were treated by a third-generation inoculant [116]. The author suggested these effects to the properties of the forage to which they are applied. Other studies have reported improvements in digestibility and steers performance fed barley silage treated with a third-generation inoculant (**Table 2**) [117, 118].

5.5. Using enzymes to enhance fiber utilization

There is increasing interest in using exogenous enzymes as a cost-effective method for improving animal productivity. The main enzyme products marketed for livestock are derived mainly from only four bacterial (*Bacillus subtilis, L. acidophilus, L. plantarum,* and *Streptococcus faecium*) and three fungal (*A. oryzae, T. reesei,* and *Saccharomyces cerevisiae*) species. Other fungal species, including *Humicola insolens* and *Thermomyces lanuginosus,* are being marketed to a lesser extent [119]. Several studies have confirmed that the addition of enzymes to feeds can increase DMI and fiber digestibility [120].

	Uninoculated	*Inoculated*	*P-value*
First generation			
DMI (kg/day)	7.13	7.05	0.40
Average daily gain (kg)	1.43	1.41	0.70
Gain: feed DM ratio	0.20	0.20	0.65
Third generation			
DMI (kg/day)	7.6	7.1	0.02
Average daily gain (kg)	1.29	1.31	065
Gain: feed DM ratio	0.17	0.19	0.02

Table 5. Effects of silage inoculants on feedlot steers performance fed whole-crop barley silage diets inoculated or uninoculated using first and third generation.

Exogenous feed enzymes with fibrolytic activities have been reported to enhance fiber digestion in the rumen [121, 122]. Most of the commercial products that have been investigated in dairy cows have had cellulases and xylanases activates, with proteases and amylases being tested in a minor number of studies. **Table 5** showed some studies that have been performed in dairy cows fed TMR supplemented with enzymes that were characterized by cellulase and/or xylanase activities. It appeared that the preparations of the current enzyme do not introduce novel enzyme activity into the rumen as they finally increase only the rate and not the extent of digestion of the cell wall [123, 124]. Beauchemin et al. reported that DMI would increase by

1.0 ± 1.3 kg/d and milk yield by 1.1 ± 1.5 kg/d with the addition of fibrolytic exogenous enzymes to dairy cow diets [125]. It is evident from the dispersion of data from the mean of the responses to the addition of enzymes fibrolytic to ruminant diets were fluctuating. Therefore, it not surprising that the use of enzyme fibrolytic products in the dairy commercial operations is not built broadly.

It is well established that the application of the exogenous enzymes before feeding is more effective when it is applied as a liquid form than as a powder. Meanwhile, spraying enzymes on the wet feed such as silage seems to be more effective than on dry feed such as hay and grain, where the wet feed is easier for enzymes to decompose the complex carbohydrates from polymers. This hydrolysis may enhance and simplify the microbial attachment, and hence reduce the lag time required for microbial colonization [126].

In high-producing dairy cattle, the stage of lactation has an important effect on the efficiency of enzyme additives. For instance, Schingoethe et al. found that the cows in early lactation responded to enzyme supplementation, but they did not detect any effect for enzymes on the cows in mid-lactation [127]. Differences in the response of early- and mid-lactation cows to enzyme supplementation were also reported in other studies [128, 129].

Enzymes that bind to feed seem to be more active, perhaps due to better resistance to proteolytic inhibition in the rumen. In general, the rumen ecosystem was found to have a minor effect on exogenous enzymes as a result of glycosylation [130]. It has also been found that nonglycosylated enzymes could sustain in the rumen and resist the proteolytic activity by ruminal microbiota, but this will be dependent on microbial sources of enzymes [131].

Due to the occurrence of internal fibrolytic enzymes yielded from the rumen bacteria, it is not easy in many cases to define the potential of exogenous enzymes to directly digest carbohydrates alone [132]. There is a synergy between the internal ruminal fibrolytic enzymes and the exogenous enzymes, where exogenous enzymes can enhance the microbial attachment to the forage fiber, here then improving fiber digestibility [133], but the mechanism by which this occurs is not known. It has been found that increasing amount of exogenous enzymes may suppress the ruminal bacteria that digest the fiber, fiber, for example, White et al. [134]. found the lower amount of exogenous enzymes enhanced the rumen bacteria attachment to fiber, in contrast, increase a number of enzymes decrease the microbial activity where exogenous enzymes have competed with ruminal bacteria enzymes for cellulose hydrogen binding sites on forage fiber. Thus, it is recommended to complement the rumen bacterial enzymes with the exogenous enzymes.

6. Conclusion

Silage contains a high content of neutral detergent fiber. Even under optimum conditions, NDF digestibility in the rumen is frequently less than 50%. Improving ruminal fiber degradability could allow cattle to consume more feed and hence increase milk yield. Selecting forage with higher NDFD could be a practical approach to increasing digestible carbohydrate and feed intake in dairy cattle. Ferulic acid-producing bacteria that are targeted at breaking the bonds between ferulic acid and hemicellulose could be the key to increasing fiber digestibility in ruminants. Addition of enzymes to feeds would increase NDFD. However, responses to feed enzymes are expected to be greatest in situations where digestible energy is the first limiting nutrient in the diet.

Acknowledgements

The Ministry of Agriculture Strategic Feed Research Chair (PY) research programs have been supported by grants from the Natural Sciences and Engineering Research Council of Canada (NSERC-Individual Discovery Grants and CRD grants), Saskatchewan Agricultural Development Fund (ADF), Ministry of Agriculture Strategic Feed Research Chair Program, Western Grain Research Foundation (WGRF), Saskatchewan Forage Network, SaskPulse, SaskCanola, SaskMilk, etc.

Author details

Basim Refat and Peiqiang Yu*

*Address all correspondence to: peiqiang.yu@usask.ca

Department of Animal and Poultry Science, College of Agriculture and Bioresources, University of Saskatchewan, Saskatoon, Canada

References

[1] Archer JA, Richardson EC, Herd RM, Arthur PF. Potential for selection to improve efficiency of feed use in beef cattle: A review. Aust J Agric Res. 1999;50(2):147–161.

[2] Guyomard H, Manceron S, Peyraud J-L. Trade in feed grains, animals, and animal products: Current trends, future prospects, and main issues. Anim Front. 2013;3:14–18.

[3] Combs DK. TTNDFD: A new approach to evaluate forages. In: Proceedings of the Cornell Nutrition Conference, Department of Animal Science, Cornell University, Ithaca, NY; 2013. p. 113–125.

[4] Van Soest PJ. Nutritional Ecology of the Ruminant, 2nd ed. 1994. Cornell University Press USA.

[5] Buxton DR, Silage science and technology, Agronomy monograph no. 42, American Society of Agronomy Inc. Madison, Wisconsin, USA, (2003): 31–93.

[6] Nikkhah A. Barley forages for modern global ruminant agriculture: A review. Rus Agri Sci. 2013;39: 206–213.

[7] Lardy G, Bauer M. Feeding Barley to Beef Cattle, North Dakota State University Extension Service, Fargo, North Dakota. 2010. Accessed May 24, 2016. Available at: http://www.ag.ndsu.edu.

[8] Beauchemin KA. Effects of dietary neutral detergent fiber concentration and alfalfa hay quality on chewing, rumen function, and milk production of dairy cows. J Dairy Sci. 1991;74(9):3140–3151.

[9] NRC. Nutrient Requirement of Dairy Cattle, 7th ed. National Research Council; 2001.

[10] Mertens DR. Creating a system for meeting the fiber requirements of dairy cows. J Dairy Sci. 1997;80:1463–1481.9241608

[11] Yansari AT, Valizadeh R, Naserian A, Christensen DA, Yu P, Shahroodi FE. Effects of alfalfa particle size and specific gravity on chewing activity, digestibility, and performance of Holstein dairy cows. J Dairy Sci. 2004;87:3912–3924

[12] Krause KM, Combs DK. Effects of forage particle size, forage source, and grain fermentability on performance and ruminal pH in midlactation cows. J Dairy Sci. 2003;86:1382–1397.

[13] Heinrichs AJ Kononoff PJ. Evaluating particle size of forages and TMRs using the new Penn state forage particle separator. Tech. Bull. DAS 96–20. Pennsylvania State University, College Agric. Sci., Cooperative Ext., University Park, PA; 2002.

[14] Yang WZ, Beauchemin KA. Physically effective fiber: Method of determination and effects on chewing, ruminal acidosis, and digestion by dairy cows. J Dairy Sci. 2006;89:2618–2633

[15] Yang WZ, Beauchemin KA. Altering physically effective fiber intake through forage proportion and particle length: Chewing and ruminal pH. J Dairy Sci. 2007;90:2826–2838

[16] Yang WZ, Beauchemin KA. Increasing physically effective fiber content of dairy cow diets through forage proportion versus forage chop length: Chewing and ruminal pH. J Dairy Sci. 2009;92:1603–1615

[17] Mertens DR. Physical and chemical characteristics of fiber affecting dairy cow performance. In: Proceedings of The Cornell Nutrition Conference, Department of Animal Science, Cornell University, Ithaca, NY; 2002. p. 124–144.

[18] Huhtanen P, Rinne M, Nousiainen J. Evaluation of the factors affecting silage intake of dairy cows: A revision of the relative silage dry-matter intake index. Animal 2007;1(5): 758–770.

[19] Waldo DR, Smith LW, Cox EL. Model of cellulose disappearance from the rumen. J Dairy Sci. 1972;55:125–129.

[20] Jung HG, Allen MS. Characteristics of plant cell walls affecting intake and digestibility of forages by ruminants. J Anim Sci. 1995;73:2774–2790.

[21] Allen MS, Mertens DR. Evaluating constraints on fiber digestion by rumen microbes. J Nutr. 1988;118:261–270.

[22] Huhtanen P, Asikainen U, Arkkila M, Jaakkola S. Cell wall digestion and passage kinetics estimated by marker and in situ methods or by rumen evacuations in cattle fed hay 2 or 18 times daily. Anim Feed Sci Technol. 2007;133(3–4):206–227.

[23] Weiss WP. Estimating the available energy content of feeds for dairy cattle. J Dairy Sci. 1998;81(3):830–839.

[24] Van Amburgh ME, Van Soest PJ, Robertson JB, Knaus WF. Corn silage neutral detergent fiber: Refining a mathematical approach for in vitro rates of digestion. Pages 99–108 in Cornell Nutrition Conference Proceedings. Cornell University, Ithaca, NY; 2003.

[25] Van Amburgh ME, Collao-Saenz EA, Higgs RJ, Ross DA, Recktenwald EB, Raffrenato E, et al. The Cornell Net Carbohydrate and Protein System: Updates to the model and evaluation of version 6.5. J Dairy Sci. 2015;98(9):6361–6380.

[26] Huhtanen P, Seppälä A, Ots M, Ahvenjärvi S, Rinne M. In vitro gas production profiles to estimate extent and effective first-order rate of neutral detergent fibre digestion in rumen. J Anim Sci. 2008;86:651–659.

[27] Combs DK. TTNDFD: A new approach to evaluate forages. In: Proceedings of the 2013 Cornell Nutrition Conference, Department of Animal Science, Cornell University, Ithaca, NY; 2013. p. 113–125.

[28] Lopes F, Ruh K, Combs DK. Validation of an approach to predict total-tract fiber digestibility using a standardized in vitro technique for different diets fed to high-producing dairy cows. J Dairy Sci. 2015;98:2596–2602.

[29] Yu P, McKinnon JJ, Christensen CR, Christensen DA. Using synchrotron-based FTIR microspectroscopy to reveal chemical features of feather protein secondary structure: Comparison with other feed protein sources. J Agric Food Chem. 2004;52(24):7353–7361.

[30] Damiran D, Yu P. Molecular basis of structural makeup of hulless barley in relation to rumen degradation kinetics and intestinal availability in dairy cattle: A novel approach. J Dairy Sci. 2011;94(10):5151–5159.

[31] Samadi, Yu P. Dry and moist heating-induced changes in protein molecular structure, protein subfraction, and nutrient profiles in soybeans. J Dairy Sci. 2011;94(12):6092–6102.

[32] Gamage IH, Jonker A, Christensen DA, Yu P. Metabolic characteristics of proteins and biomolecular spectroscopic profiles in different batches of feedstock (wheat) and their

co-products (wheat distillers dried grains with solubles) from the same bioethanol processing plant. J Dairy Sci 2012;95(11):6695–6715.

[33] Gamage IH, Yu P. Short communication: Comparison of the newly developed DVE/OEB (2010) system and the National Research Council (2001) model in modeling metabolic characteristics of proteins in dairy cattle. J Dairy Sci. 2013;96(9):5908–5913.

[34] Peng Q, Khan NA, Wang Z, Yu P. Moist and dry heating-induced changes in protein molecular structure, protein subfractions, and nutrient profiles in camelina seeds. J Dairy Sci. 2014;97(1):446–457.

[35] Peng Q, Khan NA, Wang Z, Yu P. Relationship of feeds protein structural makeup in common Prairie feeds with protein solubility, in situ ruminal degradation and intestinal digestibility. Anim Feed Sci Technol. 2014;194:58–70.

[36] Huang X, Christensen C, Yu P. Effects of conditioning temperature and time during the pelleting process on feed molecular structure, pellet durability index, and metabolic features of co-products from bio-oil processing in dairy cows. J Dairy Sci. 2015;98(7): 4869–4881.

[37] Huang X, Khan NA, Zhang X, Yu P. Effects of canola meal pellet conditioning temperature and time on ruminal and intestinal digestion, hourly effective degradation ratio, and potential nitrogen to energy synchronization in dairy cows. J Dairy Sci. 2015. ;98 (12):8836–45.

[38] Doiron K, Yu P, McKinnon JJ, Christensen DA. Heat-induced protein structure and subfractions in relation to protein degradation kinetics and intestinal availability in dairy cattle. J Dairy Sci. 2009;92(7):3319–330.

[39] Yu P. Short communication: Relationship of carbohydrate molecular spectroscopic features to carbohydrate nutrient profiles in co-products from bioethanol production. J Dairy Sci. 2012;95(4):2091–2096.

[40] Zhang X, Yu P. Differentiation of mixtures of co-product blend with barley grain based on Fourier transform infrared attenuated total reflection molecular spectroscopy: Carbohydrate molecular spectral profiles and nutritive characteristics in dairy cattle. J Dairy Sci. 2012;95(11):6624–6634.

[41] Zhang X, Yu P. Molecular basis of protein structure in combined feeds (hulless barley with bioethanol coproduct of wheat dried distillers grains with solubles) in relation to protein rumen degradation kinetics and intestinal availability in dairy cattle. J Dairy Sci. 2012;95(6):3363–3379.

[42] Becker PM, Yu P. What makes protein indigestible from tissue-related, cellular, and molecular aspects? Mol Nutr Food Res. 2013;57(10):1695–1707.

[43] Nuez-Ortín WG, Yu P. Effects of bioethanol plant and coproduct type on the metabolic characteristics of the proteins in dairy cattle. J Dairy Sci. 2010;93(8):3775–3783.

[44] Yu P, Nuez-Ortín WG. Relationship of protein molecular structure to metabolisable proteins in different types of dried distillers grains with solubles: A novel approach. Br J Nutr. 2010;104(10):1429–1437.

[45] Abeysekara S, Christensen DA, Yu P. Characterizations of structural, biochemical, and nutritive profiles in silage among cool-season corn cultivars in relation to heat units (aCHU, dCHU) with curvilinear response and multivariate analyses. J Agric Food Chem. 2013;61(50):12315–12326.

[46] Allison GG, Morris C, Hodgson E, Jones J, Kubacki M, Barraclough T, et al. Measurement of key compositional parameters in two species of energy grass by Fourier transform infrared spectroscopy. Bioresour Technol. 2009;100(24):6428–6433.

[47] Li X, Zhang Y, Hannoufa A, Yu P. Transformation with TT8 and HB12 RNAi constructs in model forage (*Medicago sativa,* Alfalfa) affects carbohydrate structure and metabolic characteristics in ruminant livestock systems. J Agric Food Chem. 2015;63(43):9590–9600.

[48] Akin DE. Chemical and biological structure in plants as related to microbial degradation of forage cell wall. In: Milligan LP, Grovum WL, Dobson A. (eds.). Control of Digestion and Metabolism in Ruminants. Prentice-Hall, Englewood Cliffs, NJ, USA; 1986. p. 139–157.

[49] Grabber JH, Ralph Hatfield RD, Quideau S. p-Hydroxyphenyl, guaiacyl, and syringyl lignins have similar inhibitory effects on wall degradability. J Agric Food Chem. 1997;45:2530–2532.

[50] Bhat MK, Hazlewood GP. Enzymology and other characteristics of cellulases and xylanases. In: Bedford MR, Partridge GG. (eds.). Enzymes in Farm Animal Nutrition. CABI Publishing, Oxon, UK; 2001. p. 11–60.

[51] Sjöström E. Wood Chemistry: Fundamentals and Applications, 2nd ed. Academic Press, Elsevier; 1993.

[52] Flint HJ, Forsberg CW. Polysaccharide degradation in the rumen: Biochemistry and genetics. In: Englehardt WV, Leonard-Marek S, Breves G, Giesecke D. (eds.). Ruminant Physiology: Digestion, Metabolism, Growth and Reproduction, Ferdinand Enke Verlag, Stuttgart; 1995. p. 43–63.

[53] Johnson DE, Ward GW, Ramsey JJ. Livestock methane: Current emissions and mitigation potential. In: Kornegay ET. (ed.). Nutrient Management of Food Animals to Enhance and Protect the Environment. Lewis Publishers, New York, NY; 1996. p. 219–234.

[54] Beever DE, Mould FL. Forage evaluation for efficient ruminant livestock production. In: Givens DI, Owen E, Axford RFE, Omed HM. (eds.). Forage Evaluation in Ruminant Nutrition. CAB International, Wallingford, UK; 2000. p. 15–42.

http://dx.doi.org/10.5772/64471

[55] Paloheimo M, Piironen J, Vehmaanperä J. Xylanases and cellulases as feed additives. In: Bedford MR, Partridge GG. (eds.). Enzymes in Farm Animal Nutrition, 2nd ed. CAB International: London, UK; 2010. p. 12–53.

[56] Humphreys JM, Chapple C. Rewriting the lignin roadmap. Curr Opin Plant Biol. 2002;5:224–229.

[57] Taboada A, Novo-Uzal E, Flores G, Loureda M, Barceló AR, Masad A, et al. Digestibility of silages in relation to their hydroxycinnamic acid content and lignin composition. J Sci Food Agric. 2010;90(7):1155–1162.

[58] Jung HG, Deetz DA. Cell wall lignification and degradability. In: Jung HG, Buxton DR, Hatfield RD, Ralph J. (eds.). Forage Cell Wall Structure and Digestibility. ASA-CSSA-SSSA: Madison, WI; 1993. p. 315.

[59] Ralph J, Quideau S, Grabber JH, Hatfield RD. Identification and synthesis of new ferulic acid dehydrodimers present in grass cell walls. J Chem Soc Perk Trans. 1994:1;3485–3498.

[60] Iiyama K, Lam TBT, Stone BA. Phenolic acid bridges between polysaccharides and lignin in wheat internodes. Phytochemistry 1990;29(3):733–737.

[61] Casler MD, Jung HG, Coblentz WK. Clonal selection for lignin and etherified ferulates in three perennial grasses. Crop Sci. 2008;48(2):424–433.

[62] Borneman, WS, Ljunghahl, LG, Hartley RD, Akin DE. Feruloyl and p-coumaryl esterases from the anaerobic fungus Neocallimastix strain MC-2: Properties and functions in plant cell wall degradation. In: Coughlan MP, Hazlewood GP. (eds.). Hemicellulose and Hemicellulases. Portland Press: Chapel Hill, NC; 1993. p. 85.

[63] Dado RG, Allen MS. Intake limitations, feeding behavior, and rumen function of cows challenged with rumen fill from dietary fiber or inert bulk. J Dairy Sci. 1995;78(1):118–133.

[64] Oba M, Allen MS. Effects of brown midrib 3 mutation in corn silage on dry matter intake and productivity of high yielding dairy cows. J Dairy Sci. 1999;82(1):135–142.

[65] Oba M, Allen MS. Effects of brown midrib 3 mutation in corn silage on productivity of dairy cows fed two concentrations of dietary neutral detergent fiber: 1. Feeding behavior and nutrient utilization. J Dairy Sci. 2000;83(6):1333–1341.

[66] Oba M, Allen MS. Evaluation of the importance of the digestibility of neutral detergent fiber from forage: Effects on dry matter intake and milk yield of dairy cows. J Dairy Sci. 1999;82(3):589–596.

[67] Jung HG, Deetz DA. Cell wall lignification and degradability. In: Jung HG, Buxton DR, Hatfield RD, Ralph J. (eds.). Forage Cell Wall Structure and Digestibility. American Society of Agronomy: Madison, WI; 1993. p. 315–346.

[68] Akin DE. Histological and physical factors affecting digestibility of forages. Agron J. 1989;81:17–25.

[69] Akin DE, Burdick D. Percentage of tissue types in tropical and temperate grass leaf blades and degradation of tissues by rumen microorganisms. Crop Sci. 1975;15:661–668.

[70] Sinclair LA, Wilkinson RG, Ferguson DMR. Effects of crop maturity and cutting height on the nutritive value of fermented whole crop wheat and milk production in dairy cows. Livest Prod Sci. 2003;81(2–3):257–269.

[71] Ahvenjärvi S, Joki-Tokola E, Vanhatalo A, Jaakkola S, Huhtanen P. Effects of replacing grass silage with barley silage in dairy cow diets. J Dairy Sci. 2006;89(5):1678–1687.

[72] Nadeau E. Effects of plant species, stage of maturity and additive on the feeding value of whole-crop cereal silage. J Sci Food Agri. 2007;87(5):789–801.

[73] Lim JM, Nestor KE, Kung L. The effect of hybrid type and dietary proportions of corn silage on the lactation performance of high-producing dairy cows. J Dairy Sci. 2015;98(2):1195–1203.

[74] Eastridge ML. Brown midrib corn silage. In: Proceedings of Tri-State Dairy Nutrition Conference. Ohio State University: Columbus, OH; 1999. p. 179–190.

[75] Gencoglu H, Shaver J. Brown midrib corn silage for lactating dairy cows: A contemporary review. Accessed May 24, 2016. Available at: http://www.uwex.edu/ces/dairynutrition/documents/BMRfeedingtrialreview2008web.pdf

[76] Weiss WP, Wyatt DJ. Effect of corn silage hybrid and metabolizable protein supply on nitrogen metabolism of lactating dairy cows. J Dairy Sci. 2006;89(5):1644–1653.

[77] Stone WC, Chase LE, Overton TR, Nestor KE. Brown midrib corn silage fed during the peripartal period increased intake and resulted in a persistent increase in milk solids yield of Holstein cows. J Dairy Sci. 2012;95(11):6665–6676.

[78] Gehman AM, Kononoff PJ, Mullins CR, Janicek BN. Evaluation of nitrogen utilization and the effects of monensin in dairy cows fed brown midrib corn silage. J Dairy Sci. 2008;91(1):288–300.

[79] Ivan SK, Grant RJ, Weakley D, Beck J. Comparison of a corn silage hybrid with high cell-wall content and digestibility with a hybrid of lower cell-wall content on performance of Holstein cows. J Dairy Sci. 2005;88(1):244–254.

[80] Taylor CC, Allen MS. Corn grain endosperm type and brown midrib 3 corn silage: Site of digestion and ruminal digestion kinetics in lactating cows. J Dairy Sci. 2005;88(4):1413–1424.

[81] Ballard CS, Thomas ED, Tsang DS, Mandebvu P, Sniffen CJ, Endres MI, et al. Effect of corn silage hybrid on dry matter yield, nutrient composition, in vitro digestion, intake by dairy heifers, and milk production by dairy cows. J Dairy Sci. 2001;84(2):442–452.

[82] Castro JJ, Bernard JK, Mullis NA, Eggleston RB. Brown midrib corn silage and Tifton 85 bermudagrass in rations for early-lactation cows. J Dairy Sci. 2010;93(5): 2143–2152.

[83] Ebling TL, Kung Jr. L. A comparison of processed conventional corn silage to unprocessed and processed brown midrib corn silage on intake, digestion, and milk production by dairy cows. J Dairy Sci. 2004;87(8):2519–2526.

[84] Thomas ED, Mandebvu P, Ballard CS, Sniffen CJ, Carter MP, Beck J. Comparison of corn silage hybrids for yield, nutrient composition, in vitro digestibility, and milk yield by dairy cows. J Dairy Sci. 2001;84(10):2217–2226.

[85] Ferraretto LF, Shaver RD. Effects of whole-plant corn silage hybrid type on intake, digestion, ruminal fermentation, and lactation performance by dairy cows through a meta-analysis. J Dairy Sci. 2015;98(4):2662–2675.

[86] Rosser CL, Gorka P, Beattie AD, Block HC, Mckinnon JJ, Lardner HA, Penner GB. Effect of maturity at harvest on yield, chemical composition, and in situ degradability for annual cereals used for swath grazing. J Anim Sci. 2013;91:3815–3826.

[87] Edmisten KL, Green JT, Mueller JP, Burns JC. Winter annual small grain forage potential. II. Quantification of nutritive characteristics of four small grain species at six growth stages. Commun Soil Sci Plant Anal. 1998;29:881–899.

[88] Bolsen KK, Berger LL. Effects of type and variety and stage of maturity on feeding values of cereal silages for lambs. J Anim Sci. 1976;42:168–174.

[89] Rustas BO, Bertilsson J, Martinsson K, Elverstedt T, Nadeau E. Intake and digestion of whole-crop barley and wheat silages by dairy heifers. J Anim Sci. 2011;89:4134–4141.

[90] Owens F. Corn genetics and animal feeding value. 66th Minnesota Nutrition Conference. St. Paul, MN; 2005. p. 20–21.

[91] Coors JG, Albrecht KA, Bures EJ. Ear-fill effects on yield and quality of silage corn. Crop Sci. 1997;37:243–247.

[92] Huang H, Faulkner DB, Singh V, Danao MC, Eckhoff SR. Effect of harvest date on yield, composition, and nutritive value of corn stover and DDGS. Am Soc Agri Biol Eng. 2012;55(5):1859–1864.

[93] Xu S, Harrison JH, Kezar W, Entrikin N, Loney KA, Riley RE. Evaluation of yield, quality, and plant composition of early-maturing corn hybrids harvested at three stages of maturity. J Prof Anim Sci. 1995;11:157–165.

[94] Curran B, Posch J. Agronomic management of silage for yield and quality: Silage cutting height. Crop Insight. 2000;10(2): 1–4. (Pioneer Hybrid International INC.)

[95] Pitzen D. Corn plant feed value. Accessed May 8, 2016. Available at: http://www.nuteam.com/cornplantart.html.

[96] Kruczyńska H, Darul K, Nowak W, Kowalik I. The chemical composition and ruminal degradability of maize silages depending on the cultivar and mowing height at harvest. J Anim Feed Sci. 2001;10(Suppl. 2):331–337.

[97] Neylon JM, Kung Jr. L. Effects of cutting height and maturity on the nutritive value of corn silage for lactating cows. J Dairy Sci. 2003;86:2163–2169.

[98] Kung Jr. L, Molder BM, Mulrooney CM, Teller RS, Schmidt RJ. The effect of silage cutting height on the nutritive value of normal corn silage hybrid compared with brown midrib corn silage fed to lactating cows. J Dairy Sci. 2008;91:1451–1457.

[99] Campos FP, Sarmento P, Nussio LG, Lugão SMB, Lima CG, Daniel JLP. Fiber monosaccharides and digestibility of milenio grass under N fertilization. Anim Feed Sci Technol. 2013;183(1–2):17–21.

[100] Fahey Jr. GC, Hussein HS. Forty years of forage quality research: Accomplishments and impact from an animal nutrition perspective. Crop Sci. 1999;39(1):4–12.

[101] Chow LO, Baron VS, Corbett R, Oba M. Effects of planting date on fiber digestibility of whole-crop barley and productivity of lactating dairy cows. J Dairy Sci. 2008;91(4): 1534–1543.

[102] Baah J, Addah W, Okine EK, McAllister TA. Effects of homolactic bacterial inoculant alone or combined with an anionic surfactant on fermentation, aerobic stability and in situ ruminal degradability of barley silage. Asian–Aust J Anim Sci. 2011;24:369–378.

[103] Addah W, Baah J, Eramus KO, McAllister TA. Silage inoculants—Are they worth the money? Accessed May 24, 2016. Available at: http://www.wcds.ca/proc/2014/Manuscripts/p%20193%20-%20208%20McAllister.pdf

[104] Reich LJ, Kung Jr. L. Effects of combining *Lactobacillus buchneri* 40788 with various lactic acid bacteria on the fermentation and aerobic stability of corn silage. Anim Feed Sci Technol. 2010;159:105–109.

[105] Driehuis F, Oude Elferink SJWH, Spoelstra SF. Anaerobic lactic acid degradation during ensilage of whole crop maize inoculated with *Lactobacillus buchneri* inhibits yeast growth and improves aerobic stability. J Appl Microbiol. 1999;87:583–594.

[106] Kendall C, Combs DK, Hoffman PC. Performance of dairy cattle fed high moisture shelled corn inoculated with *Lactobacillus buchneri*. J Dairy Sci. 2002;85(Suppl. 1):385.

[107] Ranjit NK, Taylor CC, Kung Jr. L. Effect of *Lactobacillus buchneri* 40788 on the fermentation, aerobic stability and nutritive value of maize silage. Grass Forage Sci. 2002;57:73–81.

[108] Taylor CC, Ranjit NJ, Mills JA, Neylon JM, Kung Jr. L. The effect of treating whole-plant barley with *Lactobacillus buchneri* 40788 on silage fermentation, aerobic stability, and nutritive value for dairy cows. J Dairy Sci. 2002;85:1793–1800.

[109] Kung Jr. L, Taylor CC, Lynch M, Neylon JM. The effect of treating alfalfa with *Lactobacillus buchneri* 40788 on silage fermentation, aerobic stability, and nutritive value for dairy cows. J Dairy Sci. 2003;86:336–343.

[110] Yu P, McKinnon JJ, Christensen DA. Improving the nutritional value of oat hulls for ruminant animals with pretreatment of a multienzyme cocktail: In vitro studies. J Anim Sci. 2005;83(5):1133–1141.

[111] Yu P, McKinnon JJ, Christensen DA. Hydroxycinnamic acids and ferulic acid esterase in relation to biodegradation of complex plant cell walls. Can J Anim Sci. 2005;85(3): 255–267.

[112] Nsereko VL, Smiley BK, Rutherford WM, Spielbauer AJ, Harman BR et al. Influence of inoculating forage with ferulate esterase producing lactic acid bacteria on ensilage and ruminal degradation. J Anim Sci. 2006;84(Suppl. 1):375. (Abstr.)

[113] Nsereko VL, Smiley BK, Rutherford WM, Spielbauer AJ, Harman BR, et al. Influence of a silage inoculant containing ferulate esterase producing *Lactobacillus buchneri* strain PTA6138 on aerobic stability and ruminal degradation of corn silage. J Anim Sci. 2006;84(Suppl. 1):375. (Abstr.)

[114] Krueger NA, Adesogan AT, Staples CR, Krueger WK, Dean DB, Littell RC. The potential to increase digestibility of tropical grasses with a fungal, ferulic acid esterase enzyme preparation. Anim Feed Sci Technol. 2008;145(1–4):95–108.

[115] Lynch JP, Baah J, Beauchemin KA. Conservation, fiber digestibility, and nutritive value of corn harvested at 2 cutting heights and ensiled with fibrolytic enzymes, either alone or with a ferulic acid esterase-producing inoculant. J Dairy Sci. 2015;98(2):1214–1224.

[116] Kang TW, Adesogan AT, Kim SC, Lee SS. Effects of an esterase-producing inoculant on fermentation, aerobic stability, and neutral detergent fiber digestibility of corn silage. J Dairy Sci. 2009;92(2):732–738.

[117] Addah W, Baah J, Groenewegen P, Okine EK, McAllister TA. Comparison of the fermentation characteristics, aerobic stability and nutritive value of barley and corn silages ensiled with or without a mixed bacterial inoculant. Can J Anim Sci. 2011;91(1): 133–146.

[118] Addah W, Baah J, Okine EK, McAllister TA. A third-generation esterase inoculant alters fermentation pattern and improves aerobic stability of barley silage and the efficiency of body weight gain of growing feedlot cattle. J Anim Sci. 2012;90(5):1541–1552.

[119] Muirhead S. Direct Fed Microbial, Enzyme and Forage Additive Compendium, 3rd ed. The Miller Publishing Company: Minetonka, MI. National Academy Press: Washington, DC; 1996. p. 391.

[120] Adesogan AT, Kim S-C, Arriola KG, Dean DB, Staples CR. Strategic addition of dietary fibrolytic enzymes for improved performance of lactating dairy cows. In: Proceedings

of the 18th Florida Ruminant Nutrition Symposium. University of Florida, Gainesville, FL; 2007. p. 92–110.

[121] Yang WZ, Beauchemin KA, Rode LM. Effects of an enzyme feed additive on extent of digestion and milk production of lactating dairy cows. J Dairy Sci. 1999;82:391–403.

[122] Kung L, Cohen MA, Rode LM, Treacher RJ. The effect of fibrolytic enzymes sprayed onto forages and fed in a total mixed ratio to lactating dairy cows. J Dairy Sci. 2002;85:2396–2402.

[123] Wallace RJ, Wallace SJ, McKain N, Nsereko VL, Hartnell GF. Influence of supplementary fibrolytic enzymes on the fermentation of corn and grass silages by mixed ruminal microorganisms in vitro. J Anim Sci. 2001;79:1905–1916.

[124] Jalilvand G, Odongo NE, López S, Naserian A, Valizadeh R, Shahrodi FE, Kebreab E, France J. Effects of different levels of an enzyme mixture on in vitro gas production parameters of contrasting forages. Anim Feed Sci Technol. 2008;146:289–301.

[125] Beauchemin KA, Colombatto D, Morgavi DP, Yang WZ. Use of exogenous fibrolytic enzymes to improve feed utilization by ruminants. J Anim Sci. 2003;81:E37–E47.

[126] Bowman GR, Beauchemin KA, Shelford JA. The proportion of the diet to which fibrolytic enzymes are added affects nutrient digestion by lactating dairy cows. J Dairy Sci. 2002;85:3420–3429.

[127] Schingoethe DJ, Stegeman GA, Treacher RJ. Response of lactating dairy cows to a cellulase and xylanase enzyme mixture applied to forages at the time of feeding. J Dairy Sci 1999;82(5):996–1003.

[128] Zheng W, Schingoethe DJ, Stegeman GA, Hippen AR, Treacher RJ. Determination of when during the lactation cycle to start feeding a cellulase and xylanase enzyme mixture to dairy cows. J Dairy Sci. 2000;83(10):2319–2325.

[129] Knowlton KF, McKinney JM, Cobb C. Effect of a direct-fed fibrolytic enzyme formulation on nutrient intake, partitioning, and excretion in early and late lactation Holstein cows. J Dairy Sci. 2002;85(12):3328–3335.

[130] Morgavi DP, Beauchemin KA, Nsereko VL, Rode LM, McAllister TA, Iwaasa AD, Wang Y, Yang WZ. Resistance of feed enzymes to proteolytic inactivation by rumen microorganisms and gastrointestinal proteases. J Anim Sci. 2001;79:1621–1630.

[131] Fontes CMGA, Hall J, Hirst BH, Hazlewood GP, Gilbert HJ. The resistance of cellulases and xylanases to proteolytic inactivation. Appl Microbiol Biotechnol. 1995;43:52–57.

[132] McAllister TA, Hristov AN, Beauchemin KA, Rode LM, Cheng K-J. Enzymes in ruminants diets. In: Bedford MR, Partridge GG. (eds.). Enzymes in Farm Animal Nutrition. CAB International: Wiltshire, UK; 2001.

[133] Morgavi DP, Beauchemin KA, Nsereko VL, Rode LM, Iwaasa AD, Yang WZ, McAllister TA, Wang Y. Synergy between ruminal fibrolytic enzymes and enzymes from *Trichoderma longibrachiatum*. J Dairy Sci. 2000;83:1310–1321.

[134] White BA, Mackie RI, Doerner KC. Enzymatic hydrolysis of forage cell walls. Jung HG, Buxton DR, Hatfield RD, Ralph J. Forage cell wall structure and digestibility. ASA, CSSA, SSSA, Madison, WI; 1993. p. 455–498

Section 4

Silage for Biogas Production

Chapter 8

Grass Silage for Biogas Production

Natthawud Dussadee, Yuwalee Unpaprom and Rameshprabu Ramaraj

Additional information is available at the end of the chapter

http://dx.doi.org/10.5772/64961

Abstract

Renewable energy resources of part of the Asian region are not only able to fight against climate change issues but also could contribute to economic growth, employment, and energy safety. Biogas production and use are generally regarded as a sustainable practice that can guarantee high greenhouse gas savings. Thailand is an agricultural area suitable for growing of many plants, especially annual crops that can be used as an energy crop or raw material for biogas plant. In addition, grassland biomass is suitable in numerous ways for producing energy and is the most common material for producing biogas in the present scenario. There are several types of grasses popularly growing in Thailand. Grasses are converted to silage which will be used as feedstock for anaerobic digestion. Consequently, this chapter addresses the advances in silage preparations and utilization for efficient biogas production with several digestion methods including dry and wet fermentation processes, monodigestions, and co-digestions.

Keywords: silage preparation, thai grasses, fermenters, biogas, renewable energy

1. Introduction

Agriculture is the predominant occupation of Thai people despite the constant industrial growth occurring in many parts of Thailand. In terms of agricultural lands, Thailand is also one of the largest countries in the world, especially in Asia [1]. Thailand is one of the fastest growing and energy-intensive economies in South-East Asia. Fifty percent of the total energy demand required to meet the present growth is met only through import [2]. Being a country with plenty of agricultural and energy crops, Thailand has the potential to fulfill the energy needs through biogas production [3]. Anaerobic digestion technology has emerged as one of the best technologies for the production of biogas [4]. Because of the concerns regarding energy security and environmental impact of fossil fuels, utilization of renewable energy is significantly increasing which will leads to the upgradation of living standards of people [5].

Energy crops are the type of plants cultivated as raw materials for biogas production. Agricultural lands in Thailand are well suitable for growing annual crops. Usually, temperature is warm to hot weather year-round in Thailand. The highest temperature recorded is generally during summer in the months of March till May. Most of the region receives an average rainfall of around 1100 mm. The annual crops can be used as an energy crop or raw material for biogas plant [1]. Among energy crops, grasses which belong to perennial crops are suitable due to their fastest growing rates even in infertile land, low cultivation costs, higher accessibility, consumption of whole plants, and lower environmental impacts when compared to other plants [6]. Some grass species are reported to have large amount of fibers and carbohydrates from which biogas can be produced. Many such types of grasses are popularly growing in Thailand [3, 7]. Grass substrates are converted to silage to be used as feedstock for anaerobic digestion. Energy production from silage has also attracted much interest in recent years. In the United States, perennial grasses have been stored as biomass to produce biofuels. This chapter illustrates the basic concepts of anaerobic digestion and addresses the overview of potential of grass as raw material for biogas production advance silage preparations and utilization for efficient biogas production with several digestion methods including dry and wet fermentation processes, monodigestions, and co-digestions, along with environmental impact assessment. Consequently, the aim of this chapter was to provide an overview of how to efficiently utilize the grass silage for biogas production and helpful to reduce greenhouse gas effect with environmental benefits.

2. Anaerobic digestion (AD) process

Biogas is generated from a digestion process under anaerobic conditions whose application is rapidly emerging as a viable means for providing continuous gaseous fuel and power generation. Recently, there are many countries having move towards to utilize the renewable energy especially biogas production through AD. Basically in AD, the organic materials are biologically treated in the absence of oxygen. These processes were naturally occurring through bacteria to produce "biogas." Generally biogas component is a mixture of CH_4 (40–70%), CO_2 (30–60%), and other trace gases, for example, hydrogen, hydrogen sulfide, and ammonia. The co-product from the biogas fermenter is potentially useful fertilizer in the form of a liquid or solid "digestate" [8]. For biogas production, a variety of methods are applied which can be classified in wet and dry fermentation systems.

The AD cycle represents an integrated system of a physiological process of microbial and energy metabolism, as well as the processing of raw materials under specific conditions (**Figure 1**) [4]. However, the microbial community is sensitive to variations in the operating conditions applied. AD process can be possibly integrated with other conversion processes. It could be applicable to improve their sustainability and energy balance. On the other hand, biogas system is different from other biofuels like biohydrogen, bioethanol, and biodiesel which uses only carbohydrates and lipids. Biogas is produced from all the convertible biomass macromolecules under anaerobic conditions [8, 9].

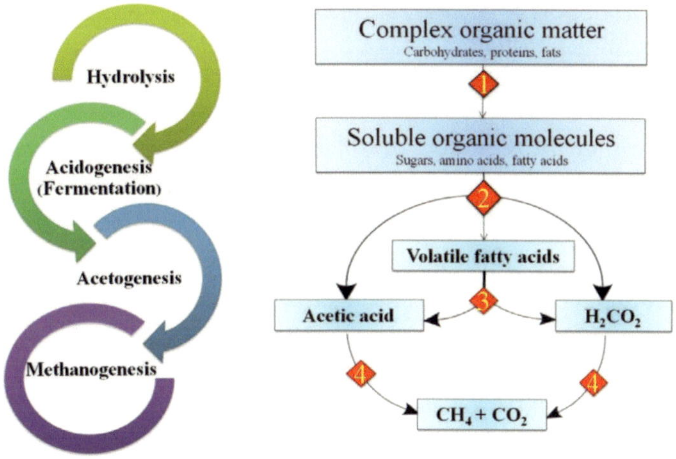

Figure 1. Flow diagram of the anaerobic digestion process.

AD is a collection of process achieved through bacteria that convert organic materials into biogas through four different stages (**Figure 1**) including hydrolysis, acidogenesis, acetogenesis, and methanogenesis [8, 9]. Organic matters are broken down step by step through these four stages towards methane production path. The complex macromolecules and components (carbohydrates, lipids, and proteins) available in organic matter are converted into simple sugars, long-chain fatty acids, and amino acids through first stage so-called hydrolysis. And second stage (acidogenesis) in turn converts these soluble micromolecules into volatile fatty acids, acetic acid, CO_2, and H_2. Third stage of acetogenesis converts the volatile fatty acids into more acetic acid, CO_2, and H_2S gas. The final stage of methanogenesis has the capability to generate methane by using the CO_2 and H_2S gas otherwise the acetic acid produced from either second or third stages [8, 9]. Thus, the AD process, if improperly managed, would become unstable and result in reduced biogas production. An overall review and assessment of AD techniques for biogas production and relevant research progress are necessary and imperative for further biogas development.

3. Grass: energy crop

Compared with other feedstocks, grass has suitable and promising characteristics as energy crop for biogas production. Because of its assurance on availability of throughout year and conservation, ensilage or haylage are indisputable. Typically, compacting to extrude sheltered air and a plastic coverage is enough for conservation of fresh grass [10]. In general, the usage of grassland as a renewable source of energy during biogas production will provide considerable quantity of environment protection, owing to the capability of grass to sequester carbon into the soil matrix. Furthermore, various socioeconomic profits are possible to achieve without harming the food industry [11].

Perennial grasses, especially C_4 grasses, are excellent candidate feedstocks for renewable energy production in support of several rationales such as high potential of dry matter yields, fast growth, and additional potential use of inputs compared to annual crops [12]. Furthermore, perennial grasses offer highest biomass yield which can be available for many harvests per year and give vital role in ecosystem services, for example, carbon sequestration in roots and soil, and to contribute the reduction of soil erosion due to massive perennial root systems that stabilize the soil. Lignin content which is negatively correlated with sugar release is lower in perennial grasses (161–192 mg g^{-1}) when compared to woody plants (157–279 mg g^{-1}) [13].

In Thailand, most of dairy cattle are grown by small-scale farmers and the grasses are used for cattle feeding. In common practice, para (*Brachiaria mutica*), ruzi (*Brachiaria ruziziensis*), guinea (*Panicum maximum*), and Napier grass (*Pennisetum purpureum*) are used in cattle feeding. Much of the prior research on candidate perennial grass biomass crops in Thailand has focused on *Brachiaria ruziziensis*, *Cynodon* sp., *Digitaria decumbens*, *Miscanthus sinensis*, *Panicum maximum*, *Paspalum atratum*, *Pennisetum polystachyon*, *Pennisetum purpureum*, *Pennisetum purpureum* × *Pennisetum americanum*, and *Vetiveria zizanioides*.

4. Thai grasses

There are many grasses already grown in Thailand that have the potential to be used as lignocellulosic feedstock for biofuel production. Several studies were suggested that wild grasses have lignocellulosic matter as new sustainable substitute raw materials for the establishment of biofuels. Many types and varieties of wild grasses are available in Thailand (**Table 1**). These grasses were potentially possible to use as a raw materials for biogas production.

Common name	Scientific name	Cultivation province	Dry matter yield (ton/ha/year)[a]
Atratum grass	*Paspalum atratum*	Chiang Mai, Lampang, Ratchaburi, or Phetchaburi	18.8
Bana grass	*Pennisetum purpureum* (Napier grass) × *Pennisetum americanum* (pearl millet)	Chiang Mai, Lampang, Ratchaburi, or Phetchaburi	7.7
	Pennisetum purpureum (Napier grass) × *Pennisetum americanum* (pearl millet)	Nakhon Ratchasima	49.1
Miscanthus grass	*Miscanthus sinensis*	Chachoengsao	N/A[b]
Mission grass	*Pennisetum polystachyon*	Nakhon Ratchasima	N/A[b]
	Pennisetum polystachyon	Phitsanulok, Phichit, Nakornsawan, Tak, Uttaradit, or Sukhothai	N/A[b]

http://dx.doi.org/10.5772/64961

Common name	Scientific name	Cultivation province	Dry matter yield (ton/ha/year)[a]
Napier grass (elephant grass)	*Pennisetum purpureum* Schum. (common)	Chiang Mai, Lampang, Ratchaburi, or Phetchaburi	7.7
	Pennisetum purpureum Schum. (common)	Nakhon Ratchasima	51.4
	Pennisetum purpureum Schum cv. Mott (Dwarf)	Chiang Mai, Lampang, Ratchaburi, or Phetchaburi	17.5
	Pennisetum purpureum Schum cv. Mott (Dwarf)	Nakhon Ratchasima	27.1
	Pennisetum purpureum Schum. cv. Kamphaeng Saen	Nakhon Ratchasima	46.3
	Pennisetum purpureum Schum. cv. King	Chiang Mai, Lampang, Ratchaburi	7.7
	Pennisetum purpureum Schum. cv. Muaklek	Nakhon Ratchasima	35.1
	Pennisetum purpureum Schum. cv. Taiwan A148	Nakhon Ratchasima	51.5
	Pennisetum purpureum Schum. cv. WrukWona		52.1
Pangola grass	*Digitaria decumbens*	Chiang Mai, Lampang, Ratchaburi	37.5
Purple guinea grass	*Panicum maximum* cv. TD 58	Chiang Mai, Lampang, Ratchaburi	18.8
	Panicum maximum cv. TD53	Nakhon Ratchasima	N/A[b]
Ruzi grass	*Brachiaria ruziziensis*	Chiang Mai, Lampang, Ratchaburi	14.1
Tifton Bermuda grass	*Cynodon nlemfuensis* cv. Tifton	Nakhon Ratchasima	58.4
Vetiver grass	*Vetiveria zizanioides* cv. Kamphaeng Phet 1	Chiang Mai, Lampang, Ratchaburi	6.5
	Vetiveria zizanioides cv. Kamphaeng Phet 2		6.0
	Vetiveria zizanioides cv. Loei		4.9
	Vetiveria zizanioides cv. Nakhon Sawan		4.2
	Vetiveria zizanioides cv. Prachuap Khiri Khan		8.5
	Vetiveria zizanioides cv. Ratchaburi		7.6
	Vetiveria zizanioides cv. Roi Et		3.5
	Vetiveria zizanioides cv. Songkhla		5.8
	Vetiveria zizanioides cv. Sri Lanka		6.4
	Vetiveria zizanioides cv. Surat Thani		5.5

[a]Banka et al. [14].
[b]Information is not available in the literature.

Table 1. Types of grasses grown in Thailand.

Brachiaria ruziziensis: Ruzi grass *(B. ruziziensis)* used mainly for domestic animals grazing. Initially, ruzi grass was native to southern African continent. It came to Thailand in 1968 from Australia. Subsequently, the grass has become popular as cattle silage because of the large production of seeds, easy to grow nature, and status as a feedstock. There are few draw backs like sensitivity to the dry climate and requirement of fertilizers [15].

Cynodon sp.: *Cynodon sp*. includes perennial grasses referred to as Bermuda grass or star grass, which are commonly grown in the topics and subtropics of the Americas, Africa, and South-East Asia [16]. Generally, they have been used for forage or as fodder for bioenergy [17]. Though Rengsirikul et al. [18] refer to Tifton grass as a type of Napier grass [18], Tifton grass is a specific breed of Bermuda grass (*Cynodon dactylon* L.) from Tifton, Georgia, USA, that was bred for its improved digestibility as a potential biofuel feedstock [17].

Digitaria decumbens: Pangola grass, scientific name *Digitaria decumbens* or *Digitaria eriantha*, is a forage grass originating from South Africa that is currently grown worldwide in the Americas, Africa, Oceania, Australia, and Asia [19]. It has been grown in Thailand since 1983 due to its success as fodder for grazing animals and its ability to grow on lands that previously cultivated rice [19].

Miscanthus sinensis: Miscanthus grass was generally called as Chinese silvergrass. Its scientific name is *Miscanthus sinensis*. Chinese silvergrass is native to eastern Asia, including Thailand. It is a perennial and clumping grass and also grown in some parts of the Americas and Europe. The grass can grow up to 2–3 meters tall [20]. Nowadays, this grass is used as cattle fodder and has been considered as a possible feedstock for biofuels.

Panicum maximum: Purple guinea grass, or *Panicum maximum* cv. Tanzania, is originally from the Ivory Coast of Africa. It is another perennial grass with a high protein content that is currently used as a feedstock for grazing animals in Thailand, having been introduced to the country in the 1980s [21].

Paspalum atratum: Atratum grass, known by its scientific name *Paspalum atratum*, is a perennial grass that can grow 1–2 meters tall. It originated in South America and is now cultivated in the Americas, South-East Asia, and Australia, generally near the equator. Though atratum grass has low drought tolerance, it is popularly grown in Thailand due to its ability to flourish during the rainy seasons and in wet soils [15].

Pennisetum polystachyon: Mission grass (*P. polystachyon*) is originally grown in tropical Africa. But for the past few decades, the grass has been spread throughout Africa, Asia, Australia, and Oceania. It can grow roughly 3 meters tall and is commonly known as a weed. The grass is a perennial and clumping grass. Mission grass is considered as an established weed that is currently not used for any specific purpose in Thailand [22].

Pennisetum purpureum: Pennisetum purpureum Schumacher, more often referred to as Napier grass or elephant grass, is a perennial grass native to Africa that has since been cultivated in tropical areas in Asia, Oceania, and the Americas. Napier grass is a hardy grass that can grow up in clumps up to seven meters in height and is particularly important as a forage and pasture grass, erosion inhibitor, mulch, and as a windbreak for other crops. Due to Napier grass's attractive qualities, such as good productivity, high yields, and drought tolerance, several types of Napier grass have already been investigated in Thailand for their potential in bioethanol conversion to bioethanol. The types of Napier grass which were already investigated include common, dwarf, Kamphaeng Saen, king, Muaklek, Taiwan, and WrukWona [6, 18].

Pennisetum purpureum × *Pennisetum americanum*: Due to the success of both Napier grass (*P. purpureum*) and pearl millet (*Pennisetum americanum*) as potential lignocellulosic feedstocks, they have been bred to create hybrids, such as bana grass [23]. Bana grass was first produced in South Africa in the 1950s and is now widely grown throughout the tropical and subtropical areas of the world [24]. Bana grass's high yield, hardiness (even when grown in harsh conditions), and its ease of harvesting have made it one of the most popular hybrids [23].

Vetiveria zizanioides: Generally, *V. zizanioides* called as vetiver grass. It is also perennial grass native to South Indian peninsula. It is used as a source of food and aromatic oils in worldwide. Furthermore, the grass has potential to apply in remediating contaminated soils, treating waste water, and reducing soil erosion [25]. Like Napier grass, vetiver grass has been examined already in Thailand as a potential source of lignocellulosic biomass for bioethanol conversion, partly due to its robustness and potential height of two meters [6, 25].

5. Napier grass

Soil fertility is generally rich in Thailand. Genus Pennisetum (including Napier grass) has been reported as the most productive tropical grasses in Thailand. Eight cultivars of Napier grass, namely Dwarf, Muaklek, Bana, Taiwan A148, Common, WrukWona, Tifton and Kamphaeng Saen, are grown in Thailand. There are several cultivars regularly grown from this genus for domestic animal feed. King Napier, Bana, WrukWona, Merkerson, and the short type (Mott dwarf) are called as common Napier. It can produce highest biomass yields more than 25 t/ha/yr dry matter when cut at 30-day intervals. In central Thailand (at Pak Chong), biomass yield was achieved at 75 t/ha/yr when cut at 60-day intervals. The scales of biomass yields demonstrated that Napier grass as a hopeful species for methane generation [18].

There is a huge awareness in the prospective utilization of Napier grass to produce ethanol in Thailand. Recently, these cultivars were selected for utilization as animal feeds, because of high leaf percentage, high nitrogen concentration, and low fiber levels. Because of its high dry matter yield, it was considered mainly as animal feed. On the contrary, for biofuels production, there is a need to get highest yield of biomass with suitability to be used either for direct combustion or for ethanol conversion. Therefore, the objectives of this paper were to quantify the yield and quality of biomass produced in different seasons by a range of Napier grass cultivars when cut at three monthly intervals throughout the year and to assess their potential as a source of energy for biofuel production in central Thailand.

In general, Rengsirikul et al. [18] confirmed that tall cultivars reach a greater length (2–4 m) than Dwarf (<1 m) with Muaklek intermediate. Furthermore, annual biomass yield was differed significantly among cultivars (**Table 2**). The tall cultivars yielded 46.3–58.4 t/ha/yr compared with 27.1 and 35.1 t/ha/yr for Dwarf and Muaklek, respectively. **Table 2** indicates that the potential of tall Napier grass cultivars to produce high biomass in Thailand to satisfy the increasing need for energy. Napier grass is tropical forage; thus, these findings can be applicable to other countries in the tropical region as well.

Cultivar	Dry matter yield (t/ha)
Dwarf	27.1
Muaklek	35.1
Bana	49.1
Taiwan A148	51.5
Common	51.4
WrukWona	52.1
Tifton	58.4
Kamphaeng Saen	46.3

Table 2. Annual dry matter (DM) yields of eight Napier grass cultivars.

6. Potential of grass silage

Several studies had been examined via grass/grass silage as feedstocks to produce biogas as a renewable energy; however, if grass is to be used as raw materials for AD for energy production, it should be converted to silage due to the presence of lignocellulosic materials [26]. Lehtomaki et al. [27] showed that AD of grass silage in batch leach bed processes has the highest methane potential when compared with other potential crops. Smyth et al. [26] compared the net energy of the grass in biomethane systems with other energy crops, and they found that grass has higher gross energy than rapeseed biodiesel and wheat ethanol systems [28]. The yields of dry matter in vetiver grass provided the yield of ethanol at 1091.84 L/ha/year, whereas the leaves of dwarf Napier grass given the maximum yield of 2720.55 L/ha/year (0.98 g/L or 0.12 g/g substrate equivalent to 30.60%) [26].

In numerous studies, grass silage has been recommended as an excellent substrate for biomethane production resulting from high-energy yields, low-energy input demand, long time storage, and usage of silage even for a whole year [29]. The higher potential of methane production from grass silage was confirmed both in batch and in semi-continuous experiments and batch leach bed processes [27]. In practice, grass silage is the most important substrate for agricultural biogas production following maize silage in Germany [30]. Though grass silage may be less energetically productive when compared to maize silage, it still offers a good energy balance and environmental advantages [31]. The key purpose of silage preparation is achieved by efficient preservation. It could keep high-energy content of a crop. And this is achieved by the combination of an anaerobic environment as well as the bacterial fermentation of sugar. The lactic acids formed in the latter progression lower the pH and avoid the proliferation of spoilage microorganisms.

Generally, the fermentation under farm conditions was not involved in a controlled process. The silage fermentation characteristics were depending on the nutrients that allow the growth of microorganisms. The fermentation is usually characterized by a low pH, high lactic acid content, and low concentrations of butyric acid and ammonia-N. Additionally, the ensiled energy is an entirely recoverable in a closed lactic acid-dominant fermentation. On the contrary, there is negligible loss of energy; the production of ethanol by yeast during fermentation is undesirable because no acidification occurs. Correspondingly, under suboptimal ensiling conditions, secondary clostridial fermentation may lead to considerable total solids and energy losses due to extensive production of CO_2 and H_2 from the fermentation of lactate and hexose sugars. If grass is to be used for energy, it must be harvested and stored, usually as silage. Silage is currently made for feeding livestock, and grass silage is mostly used as co-substrate in biogas plants based on cattle, pig, or chicken manure because of its inappropriate high nitrogen content [32, 33] of about 14% of total solids. The influence of ammonia on anaerobic digestion in terms of process inhibition was found in several literatures [34–36]. However, several authors proved that monodigestion of grass silage is possible, although both applied systems and experimental conditions differ occasionally significant.

7. Biogas from Napier grass silage

Common cultivar of Napier grass was obtained from the agriculture farm which was cultivated at Mae Taeng district, Chiang Mai, Thailand. The grass was a first cut (cut at 45-day-old mature stage). Napier grass was crushed by machine into small particles. Stored grass was pulverized into small particles (1.0 mm) before use. Proximate, ultimate, chemical composition of Napier grass is shown in **Table 3**. The grass collecting and silage preparations are shown in **Figures 2** and **3**. The experiment was carried out in the Energy Research Center, School of Renewable Energy, Maejo University, Thailand. For all experiments, Napier grass (*Pennisetum purpureum*) was used as a monosubstrate.

Property	Biomass
pH[a]	4.85
Proximate analysis (wt.%)	
Moisture[a]	77.74
Ash	3.18
Ultimate analysis (wt.%)[b]	
Carbon (C)	44.19
Hydrogen (H)	6.00
Nitrogen (N)	2.00
Oxygen (O)	43.80
Sulfur (S)	0.06

[a]As received at harvest.
[b]Dry basis; unit % by weight.

Table 3. Proximate, ultimate, chemical composition of Napier grass.

Figure 2. Grass collection and silage preparation (A) cultivation, (B) transportation of grass, (C) grass crushing machine, and (D) small particle of grass.

Figure 3. Napier grass silage.

Leachate Recirculation Digester (LBR): A prototype of 100-L dry anaerobic batch digester was employed so-called LBR system, sometimes called percolating anaerobic or dry anaerobic digester [37], and experimental setup is shown in **Figure 4**. Specification of experimental parameters and biogas measurements are listed in **Table 1**. In this design, LBR was sequentially loaded with grass biomass and mixed with residual digested solids and leachate. For all experiments, prepared grass was used as a monosubstrate. Biogas production was received

http://dx.doi.org/10.5772/64961

through improvements in the fermentation process using with Napier grass and water. Thirty kilograms of grass substrates was used in a leachate recirculation digester. The reactor working volume was 60 L.

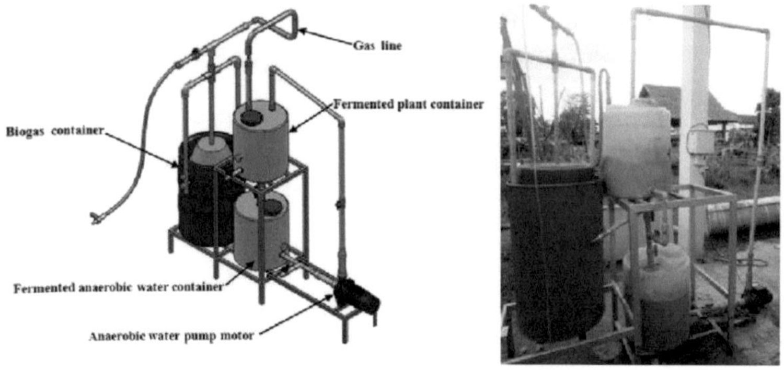

Figure 4. Dry fermentation anaerobic digestion process.

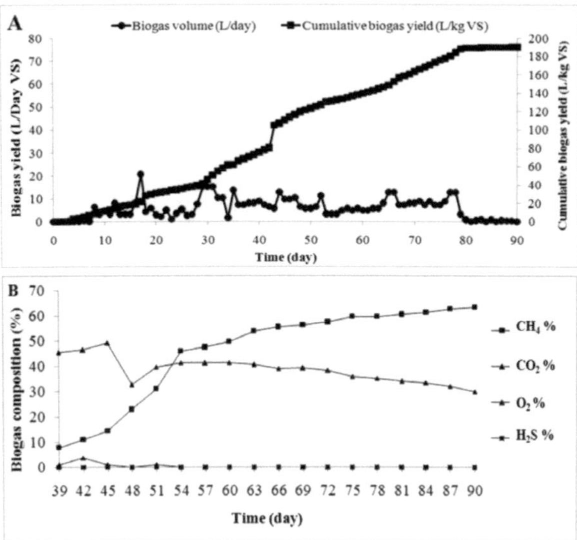

Figure 5. (A) Biogas yield (L/day VS) and cumulative biogas yield (L/kg VS) and (B) biogas compositions produced from Napier grass.

Daily total biogas production of Napier grass as monosubstrate in the reactor is given in **Figure 3**. Energy crops and crop residues can be digested either alone or in co-digestion with other materials, employing either wet or dry processes. And after 85 days, the rate of biogas production was gradually declined. The biogas was accumulated throughout study period 20.62 L/kg fresh grass or 190.25 L/kg VS is the average total amount of gas 6.87 L/day (=6870 ml/day), as shown in **Figure 5**. Bussabong et al. [38] stated the performance of the biogas production of ruzi grass (*Brachiaria ruziziensis*) as the monosubstrate had value of 244 ml/day with CSTR. This study results were demonstrated that biogas yield was 28 times higher than ruzi grass which was performed in CSTR. Batch reactors are often leach bed processes where solids are hydrolyzed by circulating leachate over a bed of organic matter. Recirculation of leachate stimulates the overall degradation owing to more efficient dispersion of inoculums, nutrients, and degradation products [27]. Accordingly, that is, main reason this study result confirmed was much higher than CSTR.

Parameter	Equipment or method
Napier grass particle size	1.00 mm
Grass substrate	30 kg
Reactor type	Leachate recirculation digester
Digesting system	Dry anaerobic digester
Volume of reactor	100 L
Used volume of reactor	60 L
Methane	ASTM D 1945
Carbon dioxide	ASTM D 1945-03
Hydrogen	ASTM D 1945-03
Hydrogen sulfide	ASTM D 5504-01
Oxygen	ASTM D1945
Sulfur	ASTM D 6667-04

Table 4. Specification of experimental parameters and biogas measurements.

Biogas composition results are presented in **Figure 5**. Biogas composition from experimental measurements starting from 39 days of the experiment showed that the initial composition of the gas as possible. This term microbial methane was generated. (Methanogenic bacteria are not in the right conditions for growth.) The pH less than 6.5 was inhibit the growth of methanogenic bacteria are composed of methane, 7.9 after 54 days, the methane production increased due to the microbial production of methane. Theoretical and measured composition of methane and biogas production is presented in **Table 4**. The biogas composition of carbon dioxide (30.10%), methane (63.50%), and 5 ppm of hydrogen sulfide was estimated from the biogas.

H_2S is commonly found in natural gas, biogas, and LPG. It is corrosive, toxic, and odorous; it can significantly damage mechanical and electrical equipment used for process control, energy generation, and heat recovery. Moreover, the combustion of H_2S results in the release of sulfur dioxide, which is a problematic environmental gas emission [39]. The usages of biogas with H_2S standard are as follows: steam and fired boilers (<1000 ppmv), steam and fired boilers (<1000 ppmv), fuel engines (<500 ppmv), motor fuels (i.e., CNG and CBG <23 ppmv), and pipe line gas (i.e., gas grid <1 ppmv) [39]. This study which verified H_2S was extremely lower (i.e., 5 ppm). Therefore, the study approach is certainly applicable for CBG (compressed biomethane gas) engine. Consequently, this study investigated the potential of Napier grass biomass as a feedstock for biogas production. This suggested that it is possible to achieve stable operation using Napier grass, as a substrate for biogas production in pilot or large-scale biogas plant in the future. It was concluded that Napier grass as energy crop can be an alternative energy resource.

7.1. Co-digestion

Recently, most of the agricultural biogas plants digest manure with the addition co-substrates to increase the content of organic material for achieving a higher gas yield [40]. For these reasons, co-digestion is commonly practiced and most recommended co-substrate was manure.

Figure 6. Wet fermentation (continuums type).

Co-digestion has been defined as the anaerobic treatment of a mixture of at least two different substrates with the aim of improving the efficiency of the anaerobic digestion process. At present, there are an increasing number of full-scale co-digestion plants treating manure and industrial organic wastes. Co-digestion of mixed substrates offers many advantages, including ecological, technological, and economic benefits, compared to digesting a single substrate. However, combining two or more different types of feed stocks requires careful selection to improve the efficiency of anaerobic digestion [40]. The main resource is represented by animal manure and slurries from cattle and pig production units as well as from poultry, fish, etc. And agricultural substrate suitable for anaerobic digestion is represented by energy crops, of which most common are grain crops, grass crops, and maize. Grass crops are among the most promising energy crops for biogas production [41].

Day	Cumulative biogas (cb-m)	Biogas component				Temp (°C)	pH
		CH_4	CO_2	O_2	H_2S (ppm)		
14	0.2618	4.9	29.6	0.9	823	31.8	5.43
15	0.6952	6.0	33.0	0.5	3877	30.5	5.65
16	1.0864	5.8	32.6	0.6	3562	29.5	5.61
17	1.4983	6.3	33.2	0.0	2325	29.2	5.45
18	2.0725	6.3	32.0	0.5	5	29.6	5.56
19	2.6462	6.9	32.0	0.1	38	30.4	5.64
20	3.2223	7.8	32.0	0.0	310	31.1	5.39
21	3.8514	8.6	32.1	0.0	423	31.7	5.48
22	4.4955	8.6	32.0	0.0	1073	31.5	5.54
23	5.1493	10.5	32.0	0.0	1458	31.9	5.42
24	5.8107	11.3	32.3	0.0	1693	30.7	5.66
25	6.8659	13.1	31.9	0.0	3715	28.5	5.68
26	7.8239	14.1	33.8	0.0	4143	29.6	5.71
27	8.4877	16.6	33.6	0.0	3972	29.9	5.74
28	9.1979	20.2	33.5	0.0	4067	28.9	5.64
29	9.7640	22.8	34.3	0.2	5345	29.7	6.08
30	10.2390	29.4	33.2	0.0	4117	28.4	6.25
31	10.8979	35.6	34.4	0.0	3623	30.1	6.08
32	11.3843	42.4	33.3	0.0	3713	30.6	6.76
33	11.8339	53.4	29.4	0.0	2522	30.6	6.78
34	12.1919	58.8	27.1	0.0	1996	27.5	6.51
35	12.7557	64.9	23.8	0.0	1592	25.1	6.85
36	13.2300	68.9	22.3	0.0	1700	24.6	6.89
37	13.5053	**70.2**	21.9	0.0	1205	25.4	6.51
38	14.1023	66.9	23.0	0.0	775	26.5	6.92
39	14.7192	62.9	26.9	0.0	1200	27.8	6.84
40	15.2051	56.9	30.4	0.0	1223	29.0	6.72

Table 5. Biogas composition and fermenter characteristic of co-digestion of Napier grass and microalgae.

In this study, we used 40-L inoculums, 1000 L of microalgae and 200 Kg of Napier silage. Microalgae was cultivated in the open pond culture, and the mesophilic anaerobic inoculum was obtained from a working mesophilic anaerobic digester at Energy Research Center, Maejo University. The inocula had a TS concentration around 296.1 ± 0.4 mg/L, with 158.5 ± 1.02 mg/L of VS. Total COD was 1241.6 mg/L, and 291.2 mg/L as $CaCO_3$ of alkalinity, 136.4

http://dx.doi.org/10.5772/64961

$mgCH_3COOH/L$ of VFA along with 6.66 of pH value. Wet fermentation (continuums type) is shown in **Figure 6**.

Gas samples were collected and analyzed, and gas components is presented in **Table 5** and **Figure 7**. The results obtained in this study suggest that co-digestion of microalgae and grass silage is a promising approach for improving biogas production. On 37 days, methane (CH_4) content was reached over 70% and CO_2 (10.05%), O_2 (21%), and H_2S 1205 ppm), which were met the standard of the Department of Energy. Efficiency criteria explained good performance throughout the study.

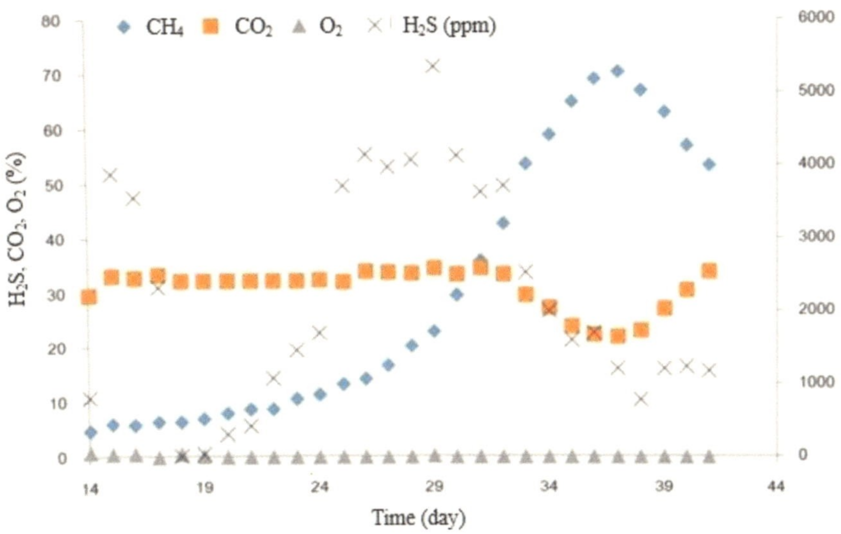

Figure 7. Biogas compositions produced from Napier grass and microalgae.

8. Conclusions

This study investigated the potential of Napier grass biomass as a feedstock for biogas production. Napier grass is fast-growing, high-yielding crops, and highly nutritious especially, so it is suitable for use as energy crops for biogas production. These results indicated that, Napier grass contains rich organic substances and these substances are suitable to use in the anaerobic fermentation process to be used to sustain microbial life and transform nutrients into biogas. Dry anaerobic digestion is a biological method used to convert organic substances into a stable product for land application without adverse environmental effects. The high content of methane (i.e., 63.50%) amount was found in total biogas from dry anaerobic fermentation in 90 days hydraulic detention time. But using with co-digestion of microalgae and Napier grass silage shows good results. In 37 days, methane content was 70%. This suggested that it is possible to achieve stable operation using Napier grass, as a substrate for biogas production with co-digestion method in pilot or large-scale biogas plant in the future. The biogas digested material is excellent source for fertilizer and it is beneficial for environmental safety and management aspects as well. It was concluded that Napier grass as energy crop can be an alternative energy resource.

Author details

Natthawud Dussadee[1,2*], Yuwalee Unpaprom[3] and Rameshprabu Ramaraj[1,2]

*Address all correspondence to: natthawud92@gmail.com

1 School of Renewable Energy, Maejo University, Chiang Mai, Thailand

2 Energy Research Center, Maejo University, Chiang Mai, Thailand

3 Program in Biotechnology, Faculty of Science, Maejo University, Chiang Mai, Thailand

References

[1] Ramaraj R, Dussadee N. Biological purification processes for biogas using algae cultures: A review. International Journal of Sustainable and Green Energy. Special Issue: Renewable Energy Applications in the Agricultural Field and Natural Resource Technology. 2015, 4(1–1): 20–32. doi:10.11648/j.ijrse.s.2015040101.14

[2] Aggarangsi P, Tippayawong N, Moran JC, Rerkkriangkrai P. Overview of livestock biogas technology development and implementation in Thailand. Energy Sustainable Development. 2013, 17: 371–377. doi:10.1016/j.esd.2013.03.004

[3] Dussadee N, Reansuwan K, Ramaraj R. Potential development of compressed bio-methane gas production from pig farms and elephant grass silage for transportation in Thailand. Bioresource Technology. 2014, 155: 438–441. doi:10.1016/j.biortech.2013.12.126

[4] Pantawong R, Chuanchai A, Thipbunrat P, Unpaprom Y. Experimental investigation of biogas production from Water Lettuce, *Pistia stratiotes* L. Emergent Life Sciences Research. 2015, 1(2): 41–46.

[5] Unpaprom Y, Intasaen O, Yongphet P, Ramaraj R. Cultivation of microalga *Botryococcus Braunii* using red Nile tilapia effluent medium for biogas production. Journal of Environmental Sciences. 2015, 3: 58–65.

[6] Wongwatanapaiboon J, Kangvansaichol K, Burapatana V, Inochanon R, Winayanuwattikun P, Yongvanich T. The potential of cellulosic ethanol production from grasses in

Thailand. Journal of Biomedicine and Biotechnology. 2012, 2012: 303748. doi: 10.1155/2012/303748

[7] Dale BE, Allen MS, Laser M, Lynd LR. Protein feeds coproduction in biomass conversion to fuels and chemicals. Biofuels. Bioproducts and Biorefining. 2009, 3(2): 219–230. doi:10.1002/bbb.132

[8] Ramaraj R, Dussadee N. Renewable energy application for organic agriculture: A review. International Journal of Sustainable and Green Energy. Special Issue: New Approaches to Renewable and Sustainable Energy. 2015, 4(1–1): 33–38. doi:10.11648/j.ijrse.s.2015040101.15

[9] Ramaraj R, Unpaprom Y, Whangchai N, Dussadee N. Culture of macroalgae *Spirogyra ellipsospora* for long-term experiments, stock maintenance and biogas production. Emergent Life Sciences Research. 2015, 1(1): 38–45.

[10] Koch K, Wichern M, Lübken M, Horn H. Mono fermentation of grass silage by means of loop reactors. Bioresource Technology. 2009, 100(23): 5934–5940. doi:10.1016/j.biortech.2009.06.020

[11] Nizami AS, Jerry D, Murphy JD. What type of digester configurations should be employed to produce biomethane from grass silage? Renewable and Sustainable Energy Reviews. 2010, 14(6): 1558–1568. doi:10.1016/j.rser.2010.02.006

[12] Karp A, Shield I. Bioenergy from plants and the sustainable yield challenge. New Phytology. 2008, 179: 15–32. doi:10.1111/j.1469-8137.2008.02432.x

[13] Fedenko JR, Erickson JE, Woodard KR, Sollenberger LE, Vendramini JMB, Gilbert RA, Helsel ZR, Peter GF. Biomass production and composition of perennial grasses grown for bioenergy in a subtropical climate across Florida, USA. BioEnergy Research. 2013, 6(3): 1082–1093.

[14] Banka A, Komolwanich T, Wongkasemjit S. Potential Thai grasses for bioethanol production. Cellulose. 2015, 22(1): 9–29. doi:10.1155/2012/303748

[15] Hare MD, Tatsapong P, Phengphet S. Herbage yield and quality of *Brachiaria* cultivars, *Paspalum atratum* and *Panicum maximum* in north-east Thailand. Tropical Grassland. 2009, 43: 65–72.

[16] Anderson W, Casler M, Baldwin B. (2008). Improvement of perennial forage species as feedstock for bioenergy. In: Vermerris W (ed.) Genetic Improvement of Bioenergy Crops. New York, NY: Springer, 347 p.

[17] Anderson WF, Dien BS, Brandon SK, Peterson JD. Assessment of bermudagrass and bunch grasses as feedstock for conversion to ethanol. Applied Biochemistry Biotechnology. 2008, 145: 13–21. doi:10.1007/s12010-007-8041-y

[18] Rengsirikul K, Ishii Y, Kangvansaichol K, Sripichitt P, Punsuvon V, Vaithanomsat P. Biomass yield, chemical composition, and potential ethanol yields of 8 cultivars of Napier grass (*Pennisetum purpureum* Schumach.) harvested 3-Monthly in Central

Thailand. Journal of Sustainable Bioenergy Systems. 2013, 3: 107–112. doi:10.4236/jsbs.2013.32015

[19] Tikam K, Phatsara C, Mikled C, Vearasilp T, Phunphiphat W, Chobtang J. Pangola grass as forage for ruminant animals: A review. Springerplus. 2013. 2: 604.

[20] Waggy MA (2011). Miscanthus sinensis [Internet]. 1999. Available from: http://www.fs.fed.us/database/feis/plants/graminoid/missin/all.html [accessed 15 February 2016].

[21] Hare MD, Phengphet S, Songsiri T, Sutin N, Stern E. Effect of cutting interval on yield and quality of two *Panicum maximum* cultivars in Thailand. Tropical Grassland. 2013, 1: 87–89.

[22] Zungsontiporn S. Global invasive plants in Thailand and its status and a case study of *Hydrocotyle umbellata* L. Plant Protection Research and Development Office, Bangkok, Thailand. In: Proceedings of the international workshop on development of database (APA5D) for biological invasion, Taichung, Taiwan, 2006.

[23] Zhang X, Gao B, Xia H. Effect of cadmium on growth, photosynthesis, mineral nutrition and metal accumulation of bana grass and vetiver grass. Ecotoxicology and Environmental Safety. 2014, 106: 102–108. doi:10.1016/j.ecoenv.2014.04.025

[24] Köster HH, Meissner HH, Coertze RJ. Variation in the production and quality of bana grass over the growing season using hand-clipped samples. South African Journal of Animal Science. 1992, 22: 31–34.

[25] Dudai N, Putievsky E, Chaimovitch D, Ben-Hur M. Growth management of vetiver (*Vetiveria zizanioides*) under Mediterranean conditions. Journal of Environmental Management. 2006, 81: 63–71. doi:10.1016/j.jenvman.2005.10.014

[26] Smyth BM, Murphy JD, O'Brien CM. What is the energy balance of grass biomethane in Ireland and other temperate northern European climates? Renewable & Sustainable Energy Reviews. 2009, 13: 2349–2360. doi:10.1016/j.rser.2009.04.003

[27] Lehtomaki A, Huttunen S, Lehtinen TM, Rintala JA. Anaerobic digestion of grass silage in batch leach bed processes for methane production. Bioresource Technology. 2008, 99: 3267–3278.

[28] Abu-Dahrieh J, Orozco A, Groom E, Rooney D. Batch and continuous biogas production from grass silage liquor. Bioresource Technology. 2011, 102(23): 10922–10928. doi: 10.1016/j.biortech.2011.09.072

[29] Prochnow A, Heiermann M, Plöchl M, Linke B, Idler C, Amon T, Hobbs PJ. Bioenergy from permanent grassland–a review: 1. Biogas. Bioresource Technology. 2009, 100(21): 4931–4944. doi:10.1016/j.biortech.2009.05.070

[30] Rösch C, Raab K, Sharka J, Stelzer V. Energy from grassland—a sustainable development? Forschungszentrum Karlsruhe, Institute for Technology Assessment and Systems Analysis, 2007 Final Report FZKA 7333, 179 p. (verified December 2008).

[31] Gerin PA, Vliegen F, Jossart JM. Energy and CO_2 balance of maize and grass as energy crops for anaerobic digestion. Bioresource Technology. 2008, 99(7): 2620–2627. doi: 10.1016/j.biortech.2007.04.049

[32] Distel RA, Didone NG, Moretto AS. Variations in chemical composition associated with tissue aging in palatable and unpalatable grasses native to central Argentina. Journal of Arid Environments. 2005, 62(2): 351–357. doi:10.1016/j.jaridenv.2004.12.001

[33] Rodriguez GA, Mandaluniz N, Flores G, Oregui LM. A gas production technique as a tool to predict organic matter digestibility of grass and maize silage. Animal Feed Science and Technology. 2005, 123–124(1): 267–276. doi:10.1016/j.anifeedsci.2005.04.035

[34] Sterling MC, Lacey RE, Engler CR, Ricke SC. Effects of ammonia nitrogen on H2 and CH4 production during anaerobic digestion of dairy cattle manure. Bioresource Technology. 2001, 77(1): 9–18. doi:10.1016/S0960-8524(00)00138-3

[35] Mignone N. Biological Inhibition and Toxicity Control in Municipal Anaerobic Digestion Facilities. Alabama Water and Pollution Control Association, USA, 2005.

[36] Strik D, Domnanovich AM, Holubar P. A pH-based control of ammonia in biogas during anaerobic digestion of artificial pig manure and maize silage. Process Biochemistry. 2006, 41(6): 1235–1238.

[37] Cysneiros D, Banks CJ, Heaven S. Anaerobic digestion of maize in coupled leach-bed and anaerobic filter reactors. Water Science and Technology. 2008, 58(7): 1505–1511. doi: 10.2166/wst.2008.518

[38] Bussabong N, Watcharenwong A, Dararat S. Biogas production from Ruzi grass in the continuous stirred tank reactor (CSTR). International Journal of Chemical, Environmental and Biological Sciences. 2013, 1(2): 348–351.

[39] Amirfakhri J, Vossoughi M, Soltanieh M. Assessment of desulfurization of natural gas by chemoautotrophic bacteria in an anaerobic baffled reactor (ABR). Chemical Engineering Progress. 2006, 45(3): 232–237. doi:10.1016/j.cep.2005.08.006

[40] Álvarez JA, Otero L, Lema JM. A methodology for optimising feed composition for anaerobic co-digestion of agro-industrial wastes. Bioresource Technology. 2010, 101: 1153–1158. doi:10.1016/j.biortech.2009.09.061

[41] Seppälä M, Laine A, Rintala J. Screening of novel plants for biogas production in northern conditions. Bioresource Technology. 2013, 139:355–362. doi:10.1016/j.biortech.2013.04.014

Chapter 9

Maize Silage as Substrate for Biogas Production

Miroslav Hutňan

Additional information is available at the end of the chapter

http://dx.doi.org/10.5772/64378

Abstract

The presented work studies the possibilities of using maize silage for biogas production in laboratory as well as in full-scale conditions. From the results of long-term operation of a mixed laboratory anaerobic reactor, it follows that processing of maize silage as a single substrate is an unstable process due to the low alkalinity of silage that has to be compensated by pH adjustment. Specific production of biogas was 0.655 m^3/kg of volatile solids. Start-up of a full-scale anaerobic reactor of a biogas plant with the volume of 2450 m^3 takes approximately 100 days. At the end of the start-up, the biogas plant reached the designed parameters—maize silage dose 6–7 t/d of total solids, the reactor load about 2.5 kg/(m^3/d) of volatile solids, the biogas production of 4200 m^3/d, electricity production of ca. 6600 kWh/d, and heat production of ca. 11,500 kWh/d. Processing of co-substrates in a biogas plant revealed both positive and negative effect on the biogas plant operation, for example, the meat and bone meal addition had a negative effect on the operation due to its high nitrogen content. Loading of crude glycerol (12.1% of the total volatile solids added) showed a positive and stabilizing effect.

Keywords: anaerobic digestion, biogas, biogas plant, maize silage, substrates for biogas production

1. Introduction

In order to replace fossil fuels by renewable energy sources, utilization of biofuels and renewable energy sources has been incorporated in the national and international legal standards and the government programs of developed countries. The EU in the Directive 2009/28/EC has defined a program of replacing 20% of total energy consumption with renewable energy sources and 10% of the consumption of liquid fuels with biofuels by 2020. One possibility of increasing the share of energy from renewable sources is biogas production from energy crops. This option has gained wide application in particular in connection with government support of the electricity price produced from biogas in many countries of the EU. In addition, growing and utilization of energetic crops for biogas production is one of the alternatives of agriculture production diversification which can significantly improve farm economics. Energy from biogas produced by anaerobic digestion of energetic crops can be utilized to improve the energetic balance of a farm as excess energy can be sold (e.g. to the electric grid). Maize in the form of silage provides high yields (10–30 t of total solids—TS per hectare [1–3]) and is thus a suitable energetic crop for biogas production. More than 17,000 biogas plants, mostly using maize silage as the main substrate, are in operation in Europe; for example, in Germany, more than 8000 biogas plants have been in operation by the end of 2015 with the plant biomass utilization of more than 52 mass% and of livestock excrements of 43 mass% [4]. The rest are industrial and agro- and food processing waste as well as municipal biowaste. Advantages of plant biomass utilization are even pronounced by the fact that 52 mass% of the total substrates processed in biogas plants result in a 79% energy production. Maize silage represents 73 mass% of the plant biomass processed in the biogas plants, while the energy represents 72% of the total energy production. Thus, in 2014, 56.88% of energy produced by biogas plants in Germany originated from maize silage [4]. Even though no precise information on the species composition of the biogas plant substrates in other countries of the European Union is available, it is clear that the main substrate is maize silage. However, only little information on its anaerobic digestion is provided in literature. Generally, it can be stated that studies on the anaerobic digestion of fresh and ensiled materials did not show any significant differences between the biogas production from these materials [5, 6]. Concerning the anaerobic digestion, the main advantage of ensiling is the conservation of plant substrates to enable biogas production for the whole year. Zauner and Kuntzel [7] present anaerobic processing of maize silage in their work achieving methane production of 0.270–0.289 m^3/kg of TS in a laboratory batch reactor. The production was somehow lower, 0.181–0.184 m^3/kg of TS, in continuous laboratory reactors. Amon et al. [2] studied the biogas production from maize and clover grass in more detail. They focused on the biogas production of various species in different stages (milk, wax, and full ripeness). Also the influence of ensiling and drying on the methane production was studied. Various species had different optimal harvesting time in different ripeness stages. Specific methane production was in the range of 0.206–0.283 Nm^3/kg of the volatile solids—VS and the methane yield was in the range of 5300 to 8530 Nm^3/ha of the VS. These results were obtained by the batch tests of the anaerobic digestion in mesophilic conditions (40°C) for 60 days. Some maize varieties showed minimum difference in the methane production considering the ripeness stage and some showed a difference of more than 25% (variety Saxxo, wax ripeness) [2]. Specific methane production of 0.282–0.419 Nm^3/kg of the VS was achieved in the work of Schittenhelm [8], who studied the effect of maize composition and ripeness stage on the methane yield. Specific methane production from hybrids with late ripeness increased with the higher date of sampling more significantly than from climatically adapted "medium-early" hybrids, which reached the maximum methane production more quickly. These results are comparable with those provided by other authors [3, 9–14]. In [13], maximum yield of 9440.6 Nm^3/ha (hybrid maize, ripeness stage FAO 400—FAO 500) and in [14], 10,401 m^3/ha (20°C and normal pressure) were obtained.

To increase the specific methane production from silage, various methods of physical, chemical, or biological pre-treatment or their combination can be used [15, 16]. However, each pre-treatment method complicates the biogas production technology and increases the operation cost. Therefore, it is necessary to always consider if a higher amount of biogas produced has a relevant effect. Considering its properties and composition, maize silage is often used in co-fermentation with other substrates, for example, with manure, sludge from wastewater plants, various plant substrates, or industrial wastes [17–21].

Here, results obtained by anaerobic digestion of maize silage as the single substrate for biogas production in laboratory as well as in full-scale conditions are presented. Start-up of a biogas plant anaerobic reactor for maize silage processing as a single substrate and operation experiences using other co-substrates is presented.

2. Anaerobic digestion of maize silage in laboratory conditions

Within the laboratory experiments on anaerobic processing of maize silage, tests of the maize silage biogas potential were carried out, and a long-term operation of a biogas production model was monitored. The aim of these tests was to determine the specific biogas production.

2.1. Test of maize silage biomethane potential

It is necessary to emphasize that the tests of biomethane potential have only an informational character and the specific methane production reached by long-term operation of anaerobic digestion of biologically degradable substrates can differ considerably. Single biomethane potential test results are significantly affected by the anaerobic sludge used as inoculum. Used anaerobic sludge is not usually adapted to the substrate degradation and during batch test the adaptation is not carried out. Therefore, the biomethane potential test provides a value lower than that obtained by long-term anaerobic digestion, when the adaptation of the anaerobic sludge to the used substrate and thus deeper anaerobic digestion of the substrate can take place. If the substrate contains a toxic or inhibitory substance, its influence might not be demonstrated during the biomethane potential test, due to the sufficient dilution of the substrate by the anaerobic sludge used for inoculation, for example, in substrates with high nitrogen or sulfur content. In long-term anaerobic digestion process, when the substrate is repeatedly supplied to the reactor, nitrogen or sulfur can accumulate in the reactor, and ammonia or sulfide inhibition of anaerobic processes can occur gradually. However, the biomethane potential test is a suitable tool for the primary evaluation of anaerobic digestion of a substrate and the possible biogas production.

Maize silage produced at the STIFI farm in Hurbanovo was used in the biomethane potential test without particle size adjustment. The particle size was given by the harvesting machine as up to 5 cm in length. Silage was made in the traditional way. Harvesting took place at the

TS of the green maize about 30%. After the cropping, green maize was compacted by bulldozer in the silage pit with dimensions of 22 m × 75 m × 5 m. For tests, the silage after two month of ensiling was used. Content of TS of the used silage represented 35% with the VS content of 95.8%. Value of pH of maize silage water leachate (100 g of silage in 400 ml of tap water) was 3.7. Anaerobically stabilized sludge from the municipal wastewater treatment plant in Devínska Nová Ves (total suspended solids— TS of 37.23 g/L and volatile suspended solids—VS of 20.74 g/L) in the volume of 0.5 l and 7 g of fresh silage was used for the tests. The sludge mixture was completed to the total volume of 1 l with tap water. To determine the biogas production from the anaerobically stabilized sludge, a blank test was done. The tests were carried out in the mesophilic temperature regime (35°C) in three repetitions. The biomethane potential results are presented in **Figure 1**.

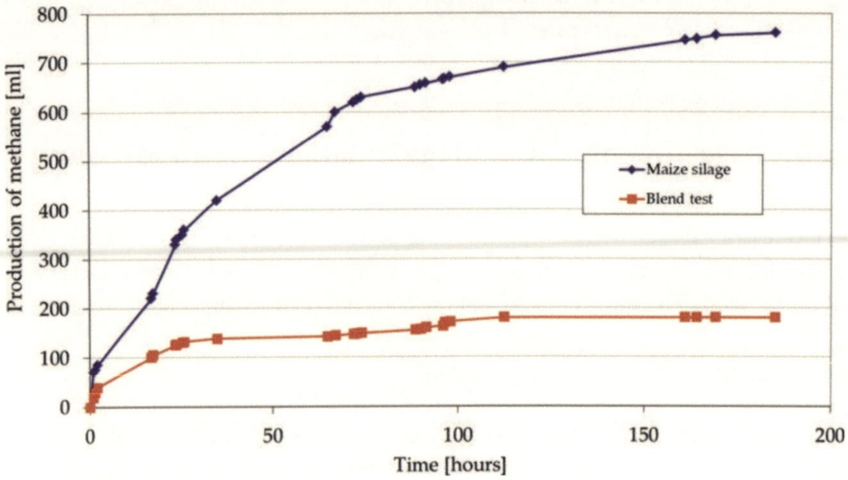

Figure 1. Test of biomethane potential of maize silage.

From the test results follows that 233 ml of methane per gram of TS (243 ml per gram of VS) respectively 0.206 Nm^3/kg of TS (0.215 Nm^3/kg of VS) were produced. It is in agreement with results provided by Amon et al. [2].

2.2. Long-term anaerobic digestion of maize silage in the laboratory conditions

Long-term maize silage processing was carried out in a mixed laboratory anaerobic reactor with the volume of 4 l. The reactor was filled to the half of its volume with the anaerobically stabilized sludge used for biomethane potential tests (TS of 37.2 g/L with VS of 55.7%) and was filled to the total volume of 4 l with tap water. Silage was processed in its raw form without any pre-treatment, that is, as taken from the silage pits in STIFI Hurbanovo, and stored at 4°C. The silage was loaded once a day into the laboratory model operated at 35°C. In the filtered samples of sludge water, parameters as chemical oxygen demand (COD), volatile fatty acids (VFA), ammonia nitrogen (NH_4-N), and pH were determined. Also the concentration of suspended solids and biogas production were monitored in the reactor. All analyses were carried out applying standard methods [22]. The analysis of VFA was done employing the method introduced by Kapp [23]. To determine the biogas composition (methane, CO_2, H_2, and H_2S), the apparatus GA 2000 Plus (Geotechnical Instruments, UK) was used.

Operation of the laboratory reactor for anaerobic digestion of maize silage started at the organic loading rate (OLR) of 1.68 kg/(m³/d). The course of loading doses and the achieved parameters of the anaerobic reactor are provided in **Table 1**. **Figure 2** shows the course of the specific biogas production per kg of added VS and the cumulative biogas production in the anaerobic reactor with gradual increase of OLR in the reactor. OLR increased from 1.68 to 6.71 kg/(m³/d)—**Table 1**. Average specific biogas production at individual OLR values was in the range of 0.195–0.655 m³ per kg of VS. Maximum specific biogas production was achieved at the OLR of 5.03 kg/(m³/d). The course of COD and VFA is shown in **Figure 3** and that of pH and NH_4-N in **Figure 4**. Instability of the processes was demonstrated by the decrease of pH and the increase of COD and VFA concentration, especially after the increase of OLR in the reactor (**Figures 2** and **3**).

Organic loading rate (VS) [kg/(m³ d)]	Day of operation	Dose of silage (raw material) [g/d]	Dose of silage (VS) [g/d]	Specific biogas production (VS) [m³/kg]
1.68	0–20	20	6.71	0.195
2.52	21–40	30	10.06	0.230
3.36	41–80	40	13.42	0.430
4.19	81–120	50	16.77	0.530
5.03	121–220	60	20.12	0.655
6.71	220–300	80	26.83	0.420

Table 1. Anaerobic processing of maize silage—operation parameters of the anaerobic reactor.

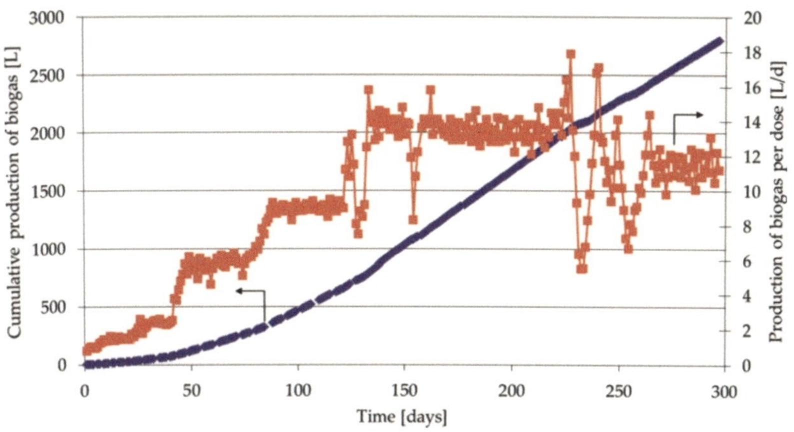

Figure 2. Specific and cumulative biogas production in the anaerobic reactor.

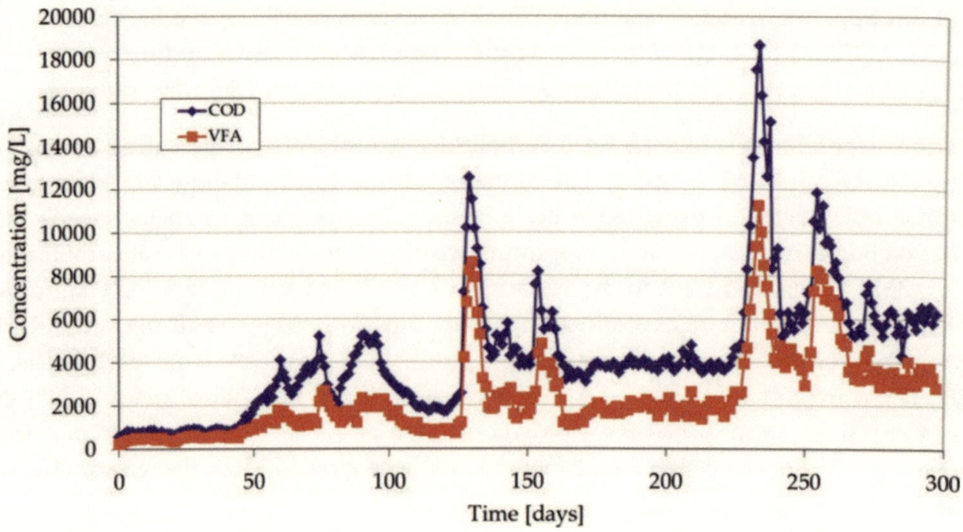

Figure 3. COD and VFA in the laboratory anaerobic reactor.

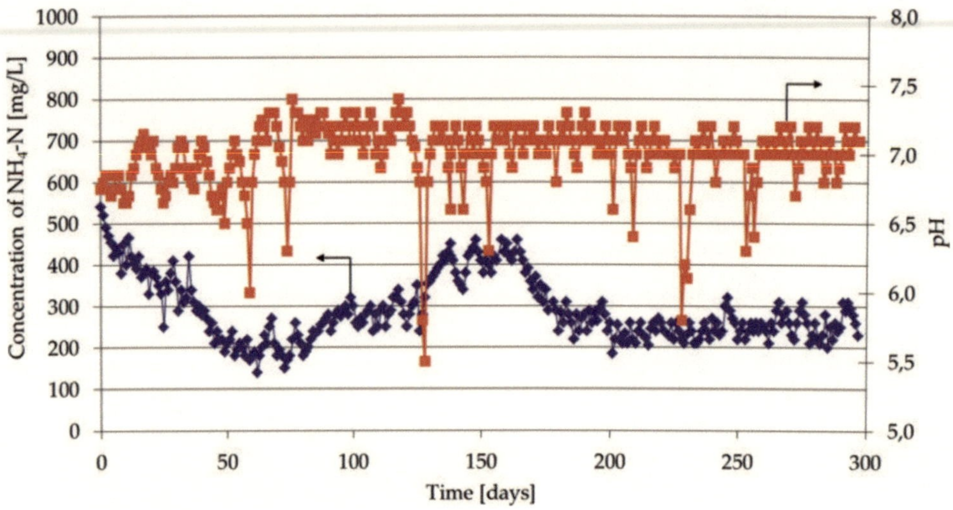

Figure 4. NH_4-N and pH in the laboratory anaerobic reactor.

Stabilization of COD and VFA took several days or even weeks depending on the destabilization degree after the silage load increase. At higher loads, the response to increased OLR was stronger and the stabilization period was longer. With the OLR increase, pH decreased below 6.5 (**Figure 3**) and sodium bicarbonate was used to adjust the pH. The pH value had to be adjusted not only after an OLR increase but throughout the anaerobic reactor operation because pH in neutral range is needed for methanogenic microorganisms. In total, ca. 6000 g of VS silage and 100 g of sodium bicarbonate were loaded into the reactor during its 300 day

http://dx.doi.org/10.5772/64378

operation. The average sodium bicarbonate consumption was 0.05 g/g of VS silage. Instability of the anaerobic processing of maize silage is related to it insufficient acid neutralizing capacity (alkalinity) due to the high C/N ratio in this substrate (30–46) [3]. Together with the carbonate buffer system ($CO_2/CO_3^{2-}/HCO_3^-$), ammonia buffer system (NH_3/NH_4^+) also has an important role in the anaerobic processes. Results of long-term anaerobic reactor operation indicate that anaerobic digestion of maize silage as the only substrate requires the presence of alkaline reagents. From a practical point of view and that of nutrients demand, loading of a co-substrate with higher content of nitrogen, for example, sewage sludge, or manure, is required. At OLR of 6.71 kg/(m^3/d), the COD and VFA values exceeded 18,000 mg/L and 11,000 mg/L, respectively. It is thus clear that at this OLR value, the system was permanently overloaded and the COD and VFA values were stabilized at 6000 mg/L and 2800 mg/L, respectively. Also the specific biogas production was considerably lower (0.420 kg/kg of VS) as that at the load of 5.03 kg/(m^3/d) (0.655 kg/kg of VS)(0.655 m^3 /kg of VS). Therefore, the optimal value of OLR has been established as 5.03 kg/(m^3/d), with the highest specific production of biogas.

During the stable operation of the reactor (days 121–220), the average concentration of suspended solids in the anaerobic reactor was 79 g/L. Daily amount of the suspended solids of excess sludge was 3.57 g. At the load of 21 g of TS silage (60 g of fresh silage with the TS content of 35%), the production of excess sludge was 0.17 g pre 1 g of TS, which corresponds to the anaerobic silage digestion degree of 83%. The content of individual biogas components is provided in **Table 2** and the parameters of anaerobic digestion of maize silage obtained from the laboratory model are summarized in **Table 3**.

Component	Value
CH_4 [%]	54.5
CO_2 [%]	45.4
H_2 [ppm]	5
H_2S [ppm]	215

Table 2. Composition of biogas produced from maize grains and maize silage.

Parameter	Value
OLR (VS) [kg/(m^3/d)]	5.03
Suspended solids in reactor [g/L]	79
Specific biogas production (35°C) [m^3/kg VS]	0.655
Specific methane production [Nm^3/kg VS]	0.316
Specific excess sludge production	0.17
Degradation of TS [%]	83.0

Table 3. Parameters of maize silage anaerobic digestion.

Considering that from 1 ha of arable land, 30 t of TS silage (VS of 95%) per annum are obtained, methane production is 9006 Nm^3/ha. For a biogas plant with produced biogas incineration in a cogeneration unit with the electric power of 1 MW (electric energy production efficiency of 35%), the daily maize silage demand represents the area of 0.77 ha. This means that the annual operation of biogas plants needs 8431.5 t of TS silage, grown on 281 ha of arable land.

Conclusions of the anaerobic digestion of maize silage in laboratory conditions:

Biomethane potential tests provided the measured specific methane production of 0.215 Nm^3/kg of VS. For long-term maize silage processing in a mixed laboratory anaerobic reactor, the measured specific methane production was 0.316 Nm^3/kg of VSS. The higher value obtained for long-term reactor operation is due to the adaptation of the anaerobic microorganisms to the maize silage substrate.

Long-term operation of the anaerobic reactor for maize silage processing as the only substrate showed significant instability caused by the low alkalinity of maize silage (high C:N ratio). To stabilize the anaerobic processes, other alkaline reagents or a co-substrate with higher content of nitrogen (sewage sludge or manure) can be used.

Daily operation of a biogas plant with biogas incinerated in a cogeneration unit with the electric power of 1 MW requires the amount of silage from an area of 0.77 ha of arable land.

3. Start-up and trial operation of biogas plant for maize silage processing

Despite the 17,000 biogas plants in EU [24], many of which use maize silage as the main substrate, only a little information on their start-up and trial operation can be found in literature. Start-up and trial operation of a biogas plant for processing of maize silage as the main substrate are described.

Technology of a biogas plant is depicted in **Figure 5**. Effective volume of the used anaerobic reactor was 2450 m^3. Two high-speed blade mixers with horizontal rotational axis and with the immersion depth and mixing direction regulation were used. Fresh silage was loaded into the reactor by means of a conveyor belt. Silage pits were located next to the anaerobic reactor; an average TS of the silage was used during the start-up, and pilot plant operation was 35%; and the expected biogas production was 4200 m^3/d. Biogas was incinerated in a cogeneration unit (ELTECO, Slovakia) with the electric power of 276 kW (electric efficiency of 32%) and the heat power of 479 kW. The reactor was operated at the temperature of 37°C and the volume of the gasholder was 80 m^3, which was assumed as sufficient at stabilized production and consumption of biogas. Also a gas boiler enabling biogas as well as a natural gas incineration with the heat power of 470 kW was included in the technology. This boiler plays an important role during the start-up of the anaerobic reactor when biogas is not available and the reactor has to be heated to the operation temperature by natural gas.

The anaerobic reactor was inoculated with aerobically stabilized sewage sludge from a brewery, which is not often used as an inoculation medium for anaerobic reactors. Normally, for the

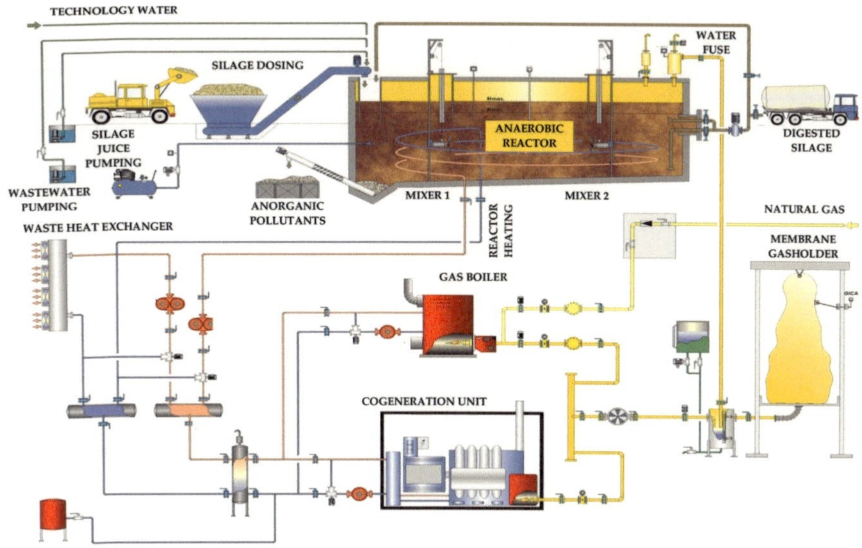

Figure 5. Diagram of biogas plant for anaerobic digestion of the maize silage.

inoculation of the anaerobic reactor the anaerobically stabilized sludge is used, but the distance to the nearest wastewater plant with anaerobically stabilized sludge was 15 km, and the required amount of sludge could not be provided. Aerobically stabilized sludge from brewery wastewater plant was available for only as far as 2 km from the biogas plant. The amount of aerobically stabilized sludge added to the anaerobic reactor for inoculation before its start-up was 1700 m^3 with an average concentration of suspended solids (SS) of 30 g/L. After the inoculation, the reactor was heated to 37°C and gradually loaded with maize silage. During the start-up of the anaerobic reactor, biogas production, pH, VFA, NH_4-N, PO_4-P, and suspended solids concentration were monitored. The course of these parameters was also monitored during the first 200 days of the pilot plant operation (**Figures 6–10**).

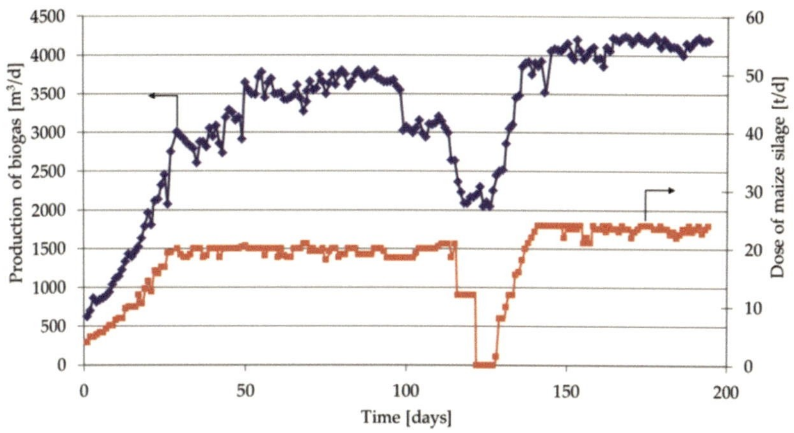

Figure 6. Course of silage dose and biogas production during the start-up of the anaerobic reactor.

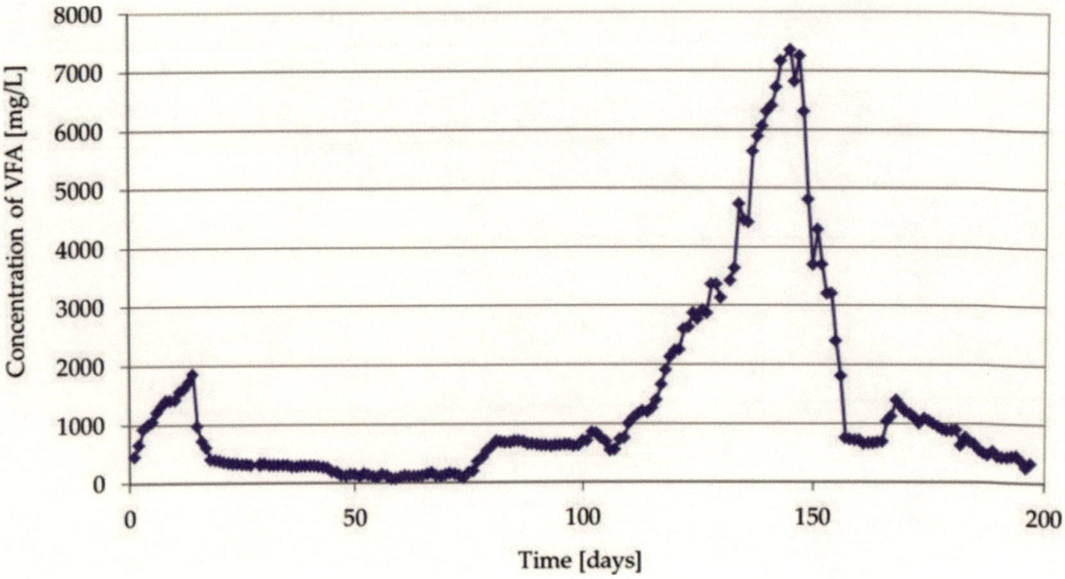

Figure 7. Concentration of VFA in filtered sludge water from the anaerobic reactor during the start-up.

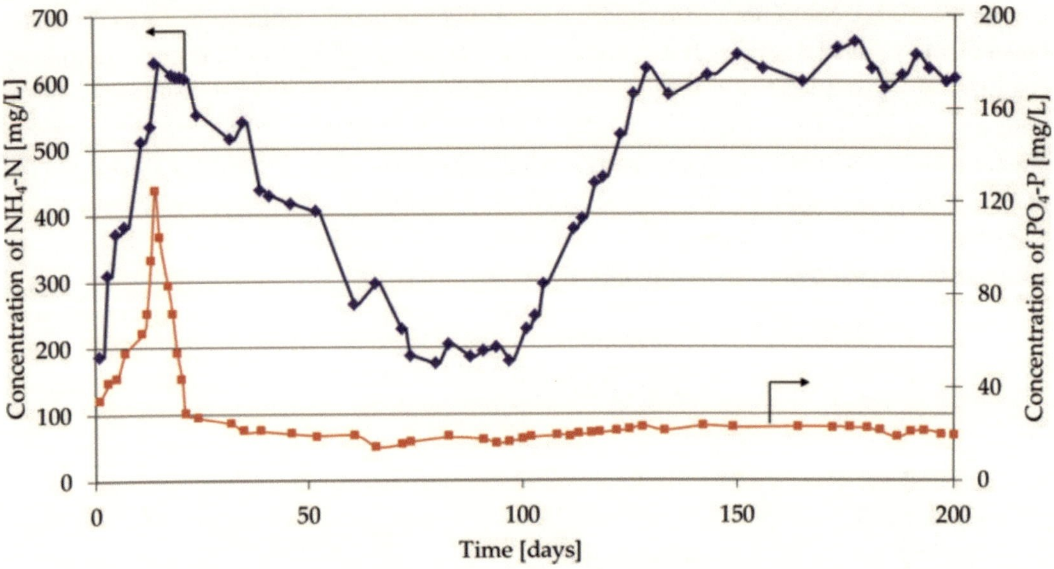

Figure 8. Concentration of NH_4-N and PO_4-P in filtered sludge water from the anaerobic during the start-up.

The silage load was gradually increased (**Figure 6**), with the starting load of 2 t/d. As it follows from **Figures 6** and **7** (biogas production and VFA concentration), anaerobic reactor operation was stable, and the biogas production was proportional to the increasing load approximately until the end of day 100. Maximum load of silage in this period was 20 t/d. This

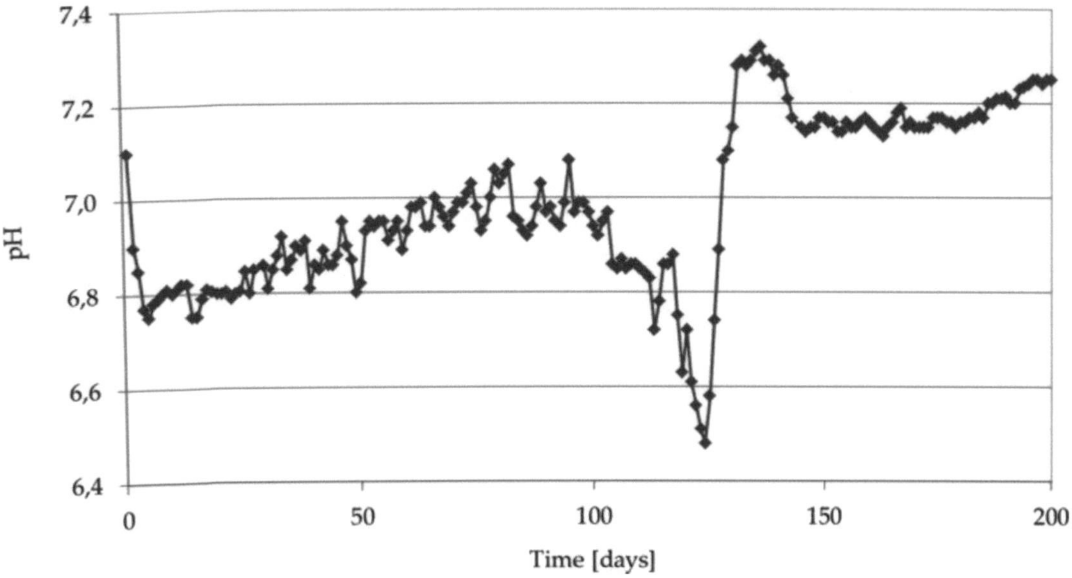

Figure 9. Course of pH in the anaerobic reactor during the start-up.

amount was divided into six parts and every 6 hours 3.33 t was dosed. Average specific biogas production between days 50 and 100 of the reactor operation was 0.726 m^3/kg of silage TS. **Figure 8** presents the course of the NH_4-N and PO_4-P concentrations.

Increase of these concentrations within the first 20 days of operation is related to the degradation of sludge used as the inoculum (similarly as for the VFA concentration—**Figure 7**). NH_4-N concentration gradually decreased to approximately 200 mg/L and that of PO_4-P to below 20 mg/L. Low concentrations of ammonia nitrogen were followed by a pH decrease (**Figure 9**) due to the low alkalinity of the silage. Values of pH below 6.5 led to methanogenesis inhibition which increased the VFA concentration above 7500 mg/L (**Figure 7**) and decreased the biogas production significantly. From day 120, it was started with dosing of aerobically stabilized sludge (the same one that was used for inoculation) to increase the NH_4-N concentration and stabilize pH. The sludge dose was 7–10 m^3/d (SS concentration of 30 g/L). As it is evident from **Figures 7–9**, the NH_4-N concentration increased to above 600 mg/L, VFA concentration decreased and pH was stabilized at around 7.2. The silage dose after stabilization of the reactor operation was increased to 24 t/d, specific biogas production reached 0.7 m^3/kg of silage TS, and the cogeneration unit worked with its 100% capacity. OLR of the anaerobic reactor was in the range of 2.3–2.7 kg/(m^3/d) and the SS concentration in sludge water of the anaerobic reactor after 200 days of operation was 60 g/L (**Figure 10**).

During the anaerobic reactor start-up, some interesting phenomena have been observed: after each silage dosing, a temporary increase in biogas production and the resulting increase in the cogeneration unit electrical power output, **Figure 11** shows the response of the electrical power output for a silage dose every 3 h (16 t per day, each dose of 2 t), with the total biogas production of 2800 m^3 and the cogeneration unit efficiency of 67% (day 140). The period of increased biogas production was ca. 1 h; the increase in biogas production showed in **Figure 11** represents 5.13% of the total biogas production per silage dose. Such an increase is related to the content of readily biodegradable organic matter in maize silage (VFA, alcohols, lower saccharides, etc.), which can vary in the range of 2.1–11.1% (**Table 4**).

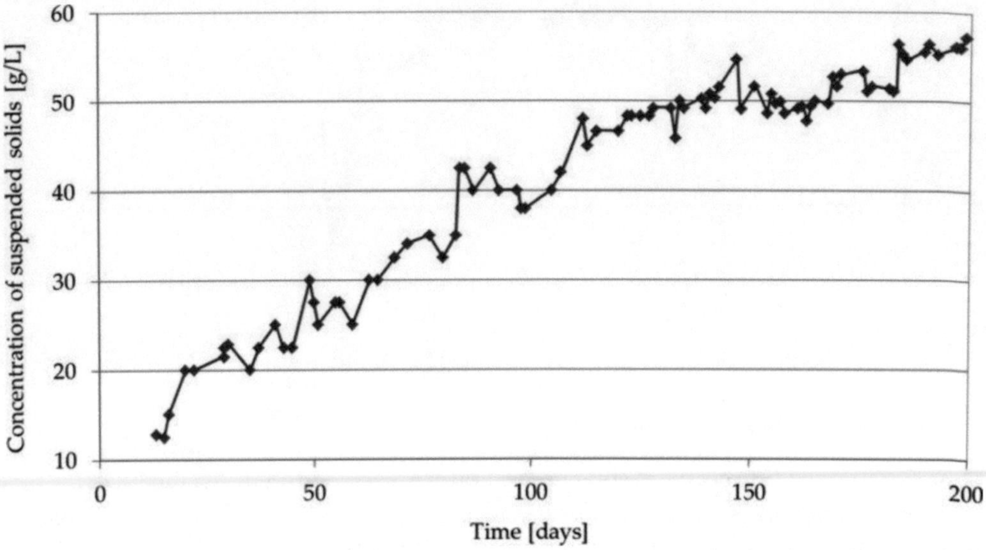

Figure 10. Concentration of suspended solids in the anaerobic reactor during the start-up.

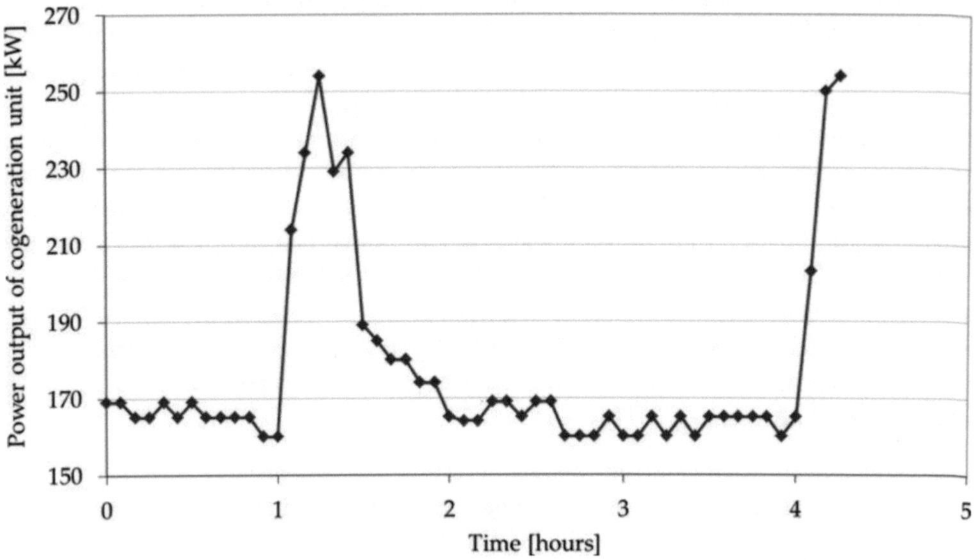

Figure 11. Increase of power output of the cogeneration unit (biogas production) after silage dosing.

TS [%]	Acid				Ethanol	Glucose	Fructose	Reference
	Lactic	Acetic	Propionic	Butyric				
39.2	6.2	2.6	0.2	<1	1.1	–	–	[25]
36.0	4.2	1.5	0.32	1.04	1.5	–	–	[26]
31.0	5.21	1.28	–	–	–	0.29	0.47	[27]
41.0	2.12	0.86	0.05	0.24	0.17	–	–	[28]
28.5	0.56	0.47	0.01	0.01	0.27	–	–	[29]

Table 4. Concentration of volatile and readily biodegradable matter in maize silage (% of TS).

Between two doses, also the quality of the biogas produced changed; average methane concentration in biogas was 54.5% and that of hydrogen sulfide was 160 ppm. In the first two hours after the loading, the methane concentration decreased by ca. 2% (from 55 to 53%), which can be explained by higher CO_2 production due to the degradation of readily biodegradable matter. More significant changes in the biogas composition were observed for a five day silage loading interruption, when the methane concentration in biogas increased from 52.8 to 65%.

In the steady state, when the full capacity of the cogeneration unit was achieved, the biogas production of 4200 m^3/d, electric energy production of ca. 6600 kWh/d and the heat production of ca. 11,500 kWh/d were provided at the daily maize silage dose of 6–7 t of TS. Electric energy was sold to the electric grid, produced heat was employed for the anaerobic reactor heating (12–13% of the produced heat), for greenhouses heating and also for drying maize grains between September and December produced on the premises as well as that produced by neighboring farmers. The digestate was stored and used if necessary as a fertilizer on the arable lands of the farm.

Conclusions from the start-up and trial operation of a biogas plant for anaerobic digestion of maize silage:

Results of the start-up of the anaerobic reactor have proved the suitability of aerobically stabilized sludge for the anaerobic reactor inoculation despite the substrate not being used for this purpose usually.

Start-up of the anaerobic reactor took approximately 100 days.

Results of the laboratory experiments were confirmed – the low alkalinity of maize silage and the need for additional substrates with higher nitrogen content to stabilize the reactor operation. In the present case, aerobically stabilized sludge from a brewery wastewater plant was used.

After ca. 150 days of the biogas plant operation, the designed parameters were stabilized. At the full capacity of the cogeneration unit, the biogas production of 4200 m^3/d, electric energy production of ca. 6600 kWh/d and heat production of ca. 11,500 kWh/d were achieved. Daily dose of silage was 24 t/d, divided into six portions every 4 hours.

4. Influence of different substrates on operation of biogas plant for anaerobic digestion of maize silage as main substrate

Raw materials used for the biogas plant differ not only in their physical properties but also in their composition. From the anaerobic digestion point of view, organic carbon and its proportion to nitrogen are more appreciated. If the organic carbon is in the form of hardly degradable matter and hydrolysis or acidification is required, the effect of substrate dosing will differ from that observed for readily biodegradable matter. If the dosing effect of readily biodegradable substrate on the anaerobic processes is strong and the biogas production increases sharply immediately after the loading, together with other changes in the reaction mixture (pH change, VFA increase, etc.), it is recommended to divide the loadings to as many as possible during the day. Dosing optimization has a positive influence on the processes not only considering the degradation but also concerning the presence of toxic or inhibitory substances.

For efficient anaerobic processes, the balanced substrates composition, especially when considering the macronutrients (nitrogen and phosphorus) content, is also an important factor. Inhibition of ongoing processes can be caused by low nitrogen or phosphorus content, or by high nitrogen content. Optimum COD/N/P ratio for the anaerobic microorganisms growth is in the range of 1000:5:1 for acidified substrate (with low biomass production), and up to 350:5:1 for unacidified substrates (with high biomass production) [29]. For materials processed in biogas plants, COD determination is quite a complex problem; therefore, the ratio of organic carbon to nitrogen (C:N ratio) is usually applied. Generally, it can be stated that for materials with high nitrogen content (blood, meat and bone meal, rapeseed meal, chicken droppings), this ratio is up to 10–15, for materials with medium nitrogen content (maize silage, cereal straw) it is up to approx. 50, and for materials with low nitrogen content (e.g. wood biomass), the C:N ratio is above 50 [30].

At very high C:N ratios, methanogenic microorganisms are not sufficiently supplied with nitrogen to assimilate (growth and propagation) and conditions for organic carbon degradation are not achieved. Or, as in case of maize silage, at low alkalinity of the substrate, pH in the reactor decreases and the process becomes instable. At low values of pH (below 6.5) growth of methanogenic microorganisms is strongly inhibited, because optimal pH for their growth is in the neutral range. However, at very low C/N ratios, nitrogen accumulates in the sludge water in its ammonia form, which can result in a pH increase and anaerobic processes inhibition by undissociated ammonia.

Considering the biomass composition, the presence of sulfur is also important; sulfur in its organic as well as inorganic form is transformed to its reduced forms, mainly to sulfides and hydrogen sulfide, by anaerobic processes. Sulfides present in the anaerobic sludge water are toxic to the methanogenic microorganisms, and hydrogen sulfide causes problems with the biogas incineration in heaters of cogeneration units.

As an example of influent of different substrates dosing on a biogas plant operation, long-term monitoring of the biogas plant in Hurbanovo using maize silage as the main substrate can be provided. Its start-up and trial operation were described above.

From the beginning of the biogas plant operation, maize silage was used as the main and often the only substrate. Maize silage composition has changed depending on different factors, for example, the maize variety used. One of the most important factors is the ripeness season when maize is harvested for ensiling [3, 8]. In **Table 5**, selected parameters of substrates significantly influencing the biogas plant operation are provided. Except for the maize silage, also meat and bone meal, molasses stillage from bioethanol production—vinasse and a by-product of biodiesel production—crude glycerol characteristics are presented.

Parameter	Maize silage	Meat and bone meal	Vinasse	Crude glycerol
pH	–	–	6.15	9.03
Chemical oxygen demand [g/g of TS]	1.22	1.32	–	–
Chemical oxygen demand [mg/L]	–	–	332,930	1,870,000
Total Kjeldahl nitrogen [% of TS]	0.88	7.90	–	–
Total Kjeldahl nitrogen [mg/L]	–	–	19,254	–
NH_4-N [mg/L]	–	–	3390	–
Total nitrogen [% of TS]	2.69	8.85	–	–
Total nitrogen [mg/L]	–	–	–	1690
Total phosphorus [mg/L]	–	–	–	192
PO_4-P [mg/L]	–	–	835	–
C:N	17.6	4.42	–	–
Dissolved anorganic salts [mg/L]	–	–	–	5150
Density [g/L]	–	–	–	1080
Lactic acid [%]	1.85	–	–	–
Acetic acid [%]	1.77	–	–	–
TS [%]	30.8	72.5	45.9	–
VS of TS [%]	94.1	79.9	71.8	–

Table 5. Characteristics of the used substrates.

During the biogas plant operation, other substrates were also processed, for example, rye silage and a mixture of oat and peas, which however did not significantly affect the reactor operation.

Co-substrate loading in the biogas plant had two main reasons: insufficient amount of the main substrate—maize silage, especially in spring; and stabilization of the anaerobic reactor operation, for example, in case of the brewery wastewater plant sludge. The course of various substrates loading is presented in **Figure 12**.

Loading of various substrates is discussed in relationship with the biogas production and pH changes. The course of biogas production in a biogas plant from the start of its operation is presented in **Figure 13** and the pH values are shown in **Figure 14**.

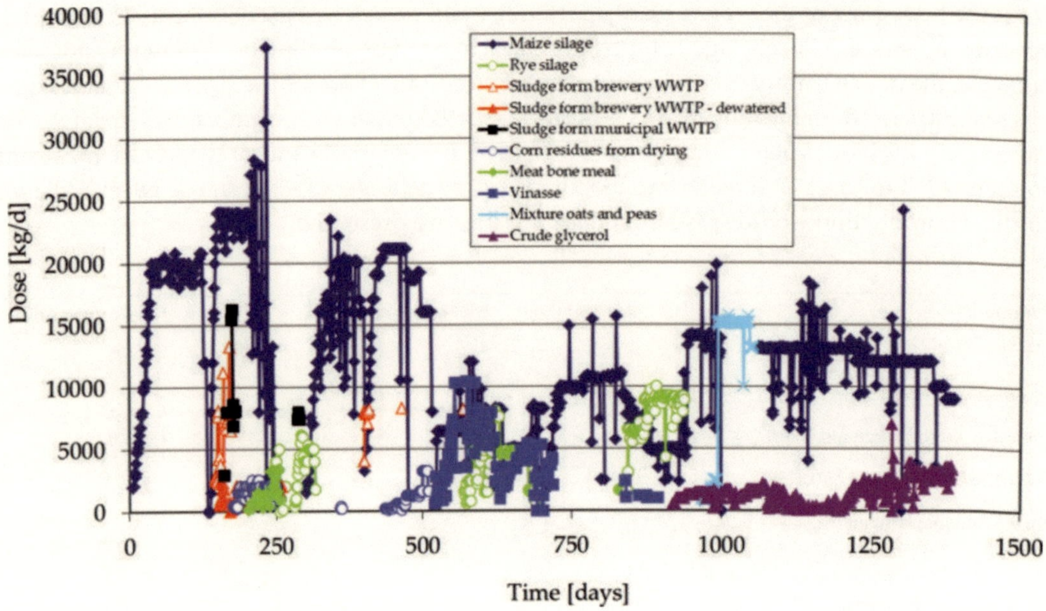

Figure 12. Course of dosing of various substrates in the biogas plant.

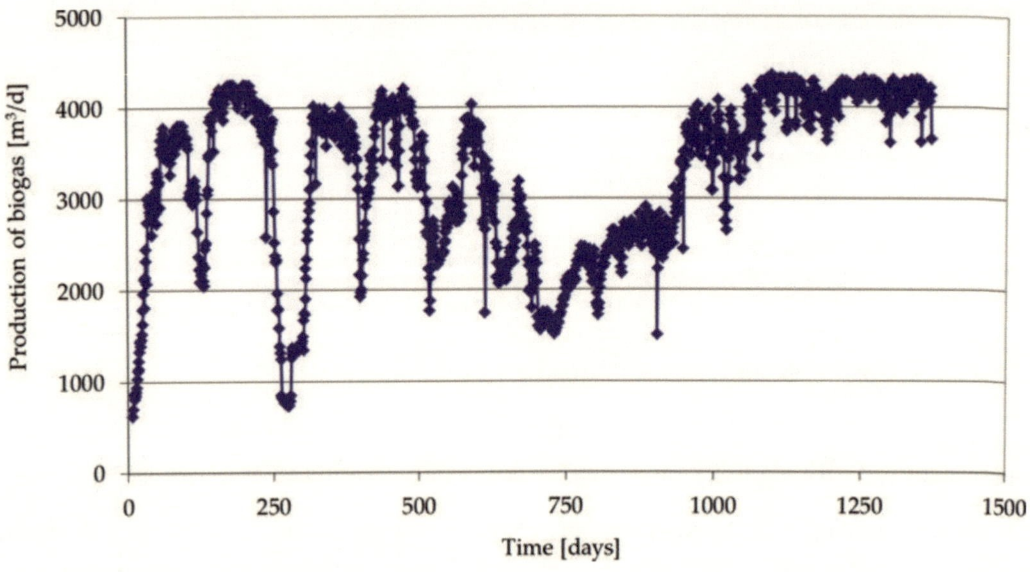

Figure 13. Biogas production in the biogas plant.

Start-up and trial operation were discussed earlier. At this period of the anaerobic reactor operation, the disadvantage of the low alkalinity of maize silage processed as the only substrate was demonstrated. This disadvantage was suppressed by brewery sludge loading to the

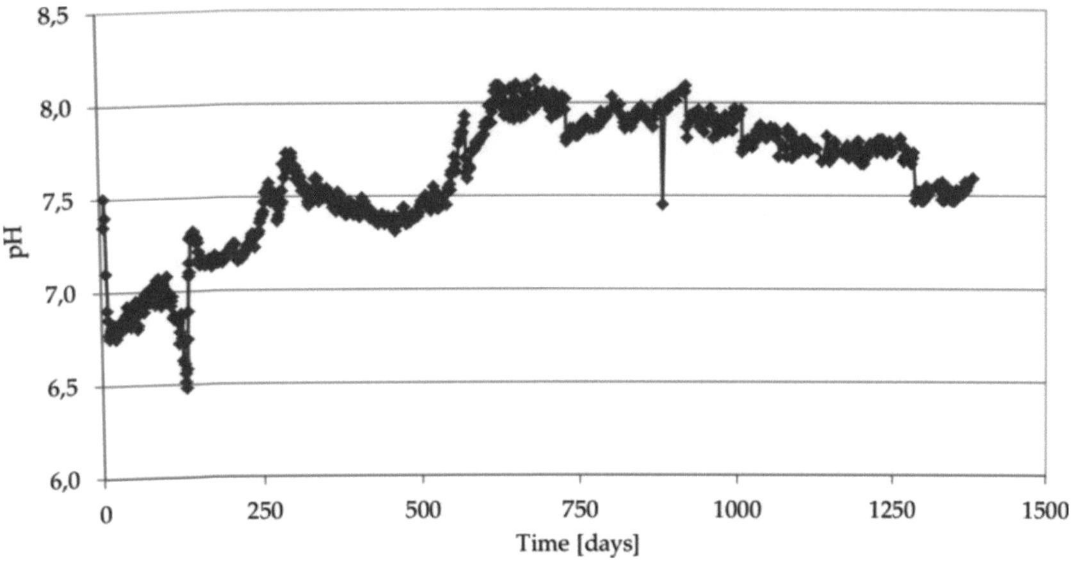

Figure 14. Course of pH values in the anaerobic reactor of the biogas plant.

reactor. The main effect of sludge dosing was the increase of N in the system. It is possible, that the other effects of sludge dosing was introducing micronutrients and minerals, but from **Figures 8** and **9** it is obvious that main effect to pH stabilization is the increase of NH_4-N concentration.

From day 203 of the operation, meat and bone meal was dosed to the reactor, first to increase nitrogen concentration, then also to compensate the lack of maize silage. When the maize silage was not available and only low quality rye silage could be used, the average dose of this silage was 2 t and up to 3.2 t of meat and bone meal. Between days 180 and 250 of the operation, the biogas production achieved its maximum (designed) values. Long-term high loads of meat and bone meal (between days 203–255) had, however, a negative impact on the anaerobic reactor operation; therefore, dosing of this substrate was in day 255 stopped. Meat and bone meal is a substrate with low C:N ratio, of 4.42 (**Table 5**), and thus with high nitrogen content, which resulted in the increase in ammonia nitrogen concentration in the sludge water to above 2800 mg/L and that of pH to above 7.5; the process was probably inhibited by undissociated ammonia in the sludge water. Volatile fatty acids concentration was higher than 7500 mg/L (**Figure 15**) and the biogas production decreased (**Figure 13**). Also the biogas quality was lower, H_2S concentration increased from values below 200 ppm to values above 1800 ppm. Loading to the anaerobic reactor was interrupted for two weeks and after the volatile fatty acids concentration decreased below 6000 mg/L, rye silage and maize splits loadings were slowly resumed. Following the initiation of maize silage dosing from the new harvest in about day 300, the reactor operation gradually stabilized and the biogas production almost reached its maximum values. Although the following reactor operation was not quite stable, such severe conditions as after meat and bone meal dosing did not occur.

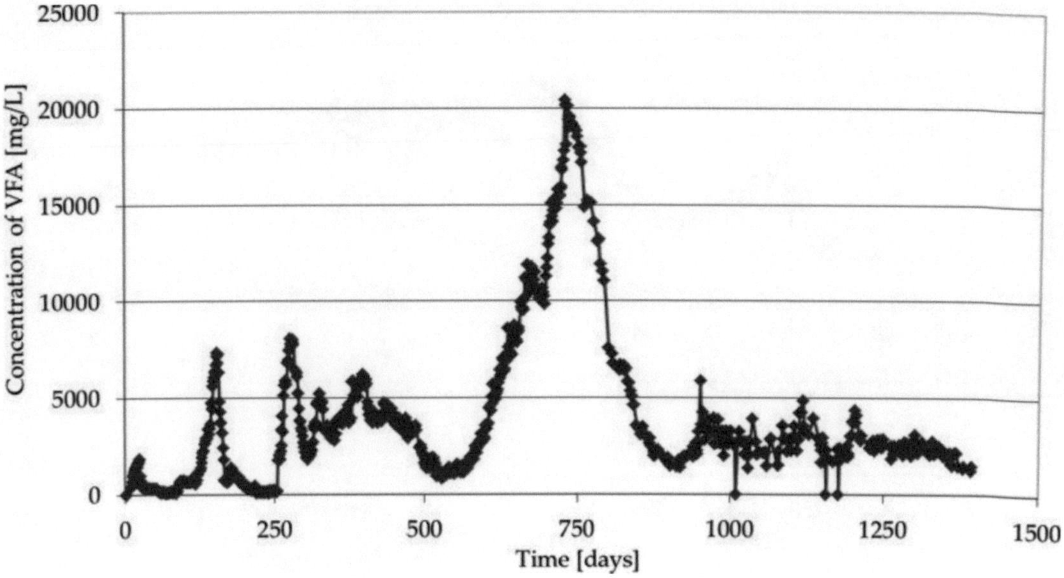

Figure 15. Concentration of VFA in filtrated sludge water from the anaerobic reactor in the biogas plant.

Another interesting substrate used in the biogas plant was molasses stillage from ethanol production, also called vinasse. At COD of 333,000 mg/L, the TKN concentration was 19,250 mg/L (**Table 5**), which is quite high. After the meat and bone meal dosing was stopped, the NH_4-N concentration in sludge water decreased below 1600 mg/L; however, it increased to almost 4500 mg/L after ca. 200 days of molasses residue loading (between days 500 and 700). Also the volatile fatty acids concentration increased significantly (**Figure 15**) and pH reached values of above eight (**Figure 14**). These changes were not abrupt but gradual and the anaerobic biomass adapted to these new conditions; therefore, the changes had no significant effect on the biogas production. Lower biogas production in this time period was caused by the lack of maize silage, as only one third or one half of the designed amount was loaded. After the molasses stillage loadings were stopped, the volatile fatty acids concentration decreased again (**Figure 15**). In the following season when maize silage was lacking, rye silage or a mixture of oat and peas was used.

Considering the anaerobic reactor operation stability, crude glycerol seems to be a promising co-substrate; it was used for more than two years in the biogas plant. It is a by-product of biofuel production; some of its characteristics are listed in **Table 5**. As it can be seen from the biogas production (**Figure 13**), pH (**Figure 14**), and volatile fatty acids concentration (**Figure 15**), the use of crude glycerol as a co-substrate with maize silage resulted in the anaerobic reactor stabilization.

To evaluate the specific biogas production from crude glycerol and its contribution to the total biogas production, a stable biogas plant operation period of 141 days was chosen, when only maize silage and crude glycerol were loaded to the anaerobic reactor. The average daily biogas production achieved was 4091.4 m^3, at the daily silage loading of 5280.4 kg TS (TS in the silage

http://dx.doi.org/10.5772/64378

used was 42.47%) and the daily crude glycerol loading of 683.7 kg. To determine the specific biogas production from crude glycerol, the value of the specific biogas production from maize silage (0.66 m^3 per 1 kg of the maize silage TS, obtained when maize silage was the only substrate) was employed. The average daily amount of biogas produced from maize silage was calculated as 3485 m^3 and the average daily amount of biogas produced from crude glycerol was 606.4 m^3. The specific biogas production per 1 kg of crude glycerol was 0.887 m^3, which corresponds to the specific biogas production of 0.512 m^3/kg of COD and is in agreement with the results presented in [31].

Biogas produced from crude glycerol represented 14.82% of the total biogas production, while 11.46% of the total TS, 12.1% of VS and only 5.21% of the total mass of the raw materials loaded. At the electrical power output of the cogeneration unit of 300 kW (electrical power output of the cogeneration unit was increased from 276 kW to 300 kW after an agreement with the producer considering the operation experiences), the daily electric energy production from crude glycerol was 1067 kWh and almost 15% of silage were saved.

Amon et al. [32] studied the influence of various loading doses of crude glycerol on the anaerobic digestion of pig manure, maize silage and maize corns. Co-fermentation effect was observed. It means that methane yield of the basic mixture supplemented with glycerol was higher than the combined methane yields of both substrates if digested separately. The co-fermentation effect was especially high with glycerin additions of 3–6%. They recommend the glycerol content of maximum 6% for a stable reactor operation.

To complete the biogas plant monitoring results obtained for various co-substrates, the course of suspended solids concentration in the anaerobic reactor is provided in **Figure 16**, which shows a gradual increase of this concentration in the period of more than two years. After crude glycerol started to be added to the reactor, the concentration of suspended solids slightly decreased, which had a positive effect on the reactor mixing.

In our case of long-term crude glycerol loading in the biogas plant, the average glycerol VS addition was 12.1% and the anaerobic digestion process was stable. However, no co-fermentation effect was observed.

Conclusions of the study of various substrates loading on the biogas plant operation:

Co-substrates used in anaerobic digestion of maize silage as the main substrate can have a positive as well as a negative effect on the biogas plant operation.

Uncontrolled meat and bone meal loading resulted in a failure of the anaerobic reactor due to the high nitrogen content in this substrate. Its processing together with the maize silage is possible; however, the loading dose has to be regulated considering the ammonia nitrogen concentration and the pH in the anaerobic reactor.

It has been proved that the inhibitory effect of ammonia nitrogen depends on also the course of its increase. While an abrupt increase of the NH_4-N concentration to approximately 2800 mg/L (loading of meat and bone meal) resulted in an inhibition of the anaerobic processes, gradual increase to almost 4500 mg/L (in case of vinasse loading) showed no negative effect on the process.

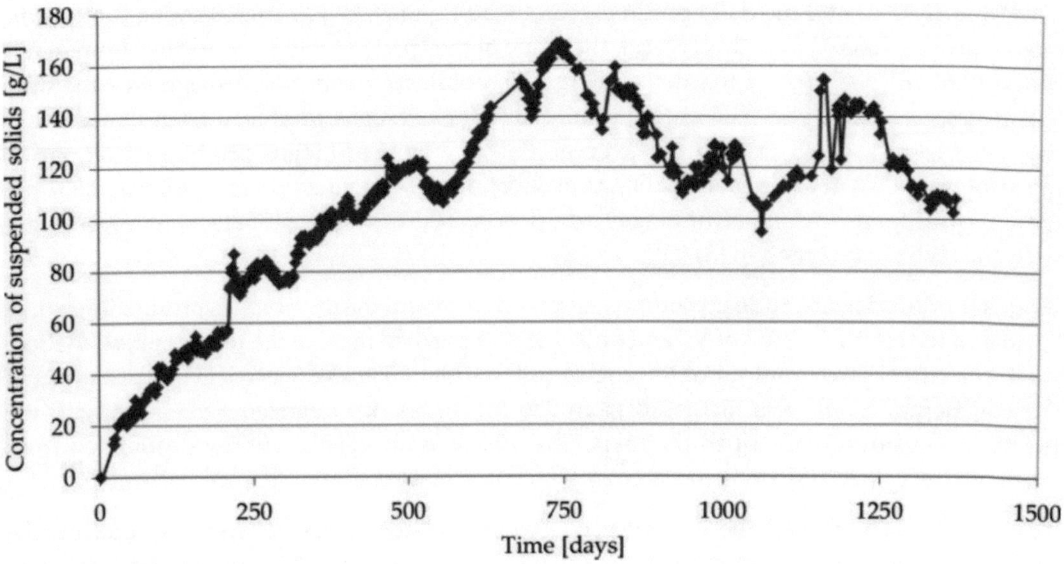

Figure 16. Concentration of undissolved substances in the anaerobic reactor in the biogas plant.

Crude glycerol loading of 12.1% of the total loaded VS had a positive and stabilizing effect on the biogas plant operation.

5. Conclusions

The obtained laboratory and full scale results showed that maize silage is a suitable substrate for anaerobic digestion and biogas production. This is confirmed also by thousands of biogas plants in Europe, which use maize silage as the main substrate. Thus, it can be concluded that maize silage as the most widely used substrate for biogas production is a fact of presence. However, biogas production from this substrate is not sustainable nor is the production of the first generation biofuels produced from food commodities (e.g. biodiesel from edible oils, or bioethanol from cereals).

The advantage of using maize silage for biogas production is because of its high yield per hectare and high specific biogas production. The main disadvantage is that maize used for biogas production cannot be used as designed in human or animal nutrition. Such competition deforms also its price and maize silage becomes a scarce commodity. Moreover, mass growing of maize as a monoculture occupies arable land for growing other crops and increases the need for fertilization and plant protection. These negative effects were also presented by the authors from the Karlsruhe Institute of Technology, Germany [33]. The main aims of their project were "Landscaping instead of monoculture" and "Grass is an alternative to silage maize in biogas production." It is evident that grass hectare yields or specific biogas production cannot compete with maize silage; however, it is a sustainable alternative to maize silage and makes biogas a surplus value of landscaping.

Another alternative is growing energetic crops which are not part of the food chain of humans or animals, for example, sorghum provides interesting hectare yields as well as specific biogas production values [34]. Other such crops include hemp (*Cannabis sativa* L.) [35] or Chinese silver grass (*Miscanthus sinensis Anderss*) [36].

Also biologically degradable waste from agriculture and industry as well as municipal waste are suitable substrates for biogas production and are an alternative to maize silage, which has been confirmed by the estimated biogas potential of these substrates in Germany [37]. While the energetic crops biogas potential was estimated to be 46.2% of the total biogas potential, the rest is obtained from various waste materials such as livestock excrements, harvesting residues and by-products of crops processing, municipal waste, sewage sludge, landscaping and industrial wastes.

Although a structure of the substrate mixture used in biogas plants is not sustainable even though its change is inevitable a long-time will pass before maize silage loses its position as the main substrate for biogas production.

Acknowledgements

This contribution is the result of the project implementation: Finalizing of the National Centre for Research and Application of Renewable Energy Sources, ITMS 26240120028, supported by the Research & Development Operational Programme funded by the ERDF.

Author details

Miroslav Hutňan

Address all correspondence to: miroslav.hutnan@stuba.sk

Faculty of Chemical and Food Technology, Institute of Chemical and Environmental Engineering, Slovak University of Technology, Bratislava, Slovak Republic

References

[1] Landbeck M., Schmidt W. Energy maize-goals, strategies and first breeding successes. CD-ROM computer file. In: Proceedings of the First International Energy Farming Congress, Papenburg, Germany, March 2–4, 2005. Kompetenzzentrum Nachwachsende Rohstoffe, Werlte, Germany, 2005. p. 2–4

[2] Amon T., Kryvoruchko V., Amon B., Moitzi G., Buga S., Lyson D.F., Hackl E., Jeremic D., Zollitsch W., Potsche E. Biogas production from the energy crops maize and clover

grass. Forschungsprojekt Nr. 1249 GZ 24.002/59-IIA1/01, Institut fur Land- und Umveltund Energietechnik. Universitat fur Bodenkultur, Wienna, Austria. 2003.

[3] Amon T., Amon B., Kryvoruchko V., Zollitsch W., Mayer K., Gruber L. Biogas production from maize and dairy cattle manure—influence of biomass composition on the methane yield. Agriculture, Ecosystems and Environment. 2007; 118:173–182. DOI: 10.1016/j.agee.2006.05.007

[4] DBFZ (Deutsches Biomasseforschungszentrum gemeinnützige GmbH). Stromerzeugung aus Biomasse (Vorhaben IIa Biomasse). Zwischenbericht. Projektnummer DBFZ, Leipzig: 2015.

[5] Zubr J. Methanogenic fermentation of fresh and ensiled plant materials. Biomass. 1986; 11:159–171. DOI: 10.1016/0144-4565(86)90064-8

[6] Gunaseelan V.N. Anaerobic digestion of biomass for methane production: a review. Biomass and Bioenergy. 1997; 13:83–114. DOI: 10.1016/S0961-9534(97)00020-2

[7] Zauner E., Küntzel U. Methane production from ensiled plant material. Biomass. 1986; 10:207–223. DOI: 10.1016/0144-4565(86)90054-5

[8] Schittenhelm S. Chemical composition and methane yield of maize hybrids with contrasting maturity. European Journal of Agronomy. 2008; 29:72–79. DOI: 10.1016/j.eja.2008.04.001

[9] Eder J., Papst C., Eder B., Krützfeldt B., Oechsner H., Mukengle M. Aktuelle Ergebneisse aus dem energiepflanzenbau-leistungspotenziale, Pflanzenbau und fruchfolgen. In: Proceeding of the First Einbecker Energiepflanzen Kolloquium, December 7–8, 2005. KWS Saat AG, Einbeck, Germany.

[10] Kaiser F., Schlattmann M., Gronauer A. Methane yield of various energy crops tests at laboratory scale and transferability to full-scale application. In: Proceedings of the Seventh International Conference on Construction, Technology, and Environment in Farm Animal Husbandry, Braunschweig, Germany, March 2–3. 2005: p. 355–360.

[11] Tatah E., Gaudchau M., Honermeier B. The impact of maize cultivar and maturity stage on dry matter, biogas and methane gas yields. Mitteilungen der Gesellschaft für Pflanzenbauwissenschaft. 2007; 19:196–197.

[12] Schumacher B., Bohmel C., Oechsner H. Which energy maize varieties when to harvest for biogas production? Landtechnik. 2006; 61:84–85.

[13] Oslaj M., Mursec B., Vindis P., Biogas production from maize hybrids. Biomass and Bioenergy. 2010; 34:1538–1545. DOI: 10.1016/j.biombioe.2010.04.016

[14] Bacenetti J., Negri M., Lovarelli D., Garcia L.R., Fiala M. Economic performances of anaerobic digestion plants: effect of maize silage energy density at increasing transport distances. Biomass and Bioenergy. 2015; 80:73–84. DOI: 10.1016/j.biombioe.2015.04.034

[15] Lehtomäki A. Biogas production from energy crops and crop residues. PhD Thesis, University of Jyväskylä, Jyväskylä, Finland; 2006.

[16] Kalač P. The required characteristics of ensiled crops used as a feedstock for biogas production: a review. Journal of Agrobiology. 2011; 28:85–96. DOI: 10.2478/v10146-011-0010-y

[17] Yangin-Gomec C., Ozturk I. Effect of maize silage addition on biomethane recovery from mesophilic co-digestion of chicken and cattle manure to suppress ammonia inhibition. Energy Conversion and Management. 2013; 71:92–100. DOI: 10.1016/j.enconman.2013.03.020

[18] Aciri S., Kocar G. The effect of adding maize silage as a co-substrate for anaerobic animal manure digestion. International Journal of Green Energy. 2015; 12:453–460. DOI: 10.1080/15435075.2013.848361

[19] Galitskaya P.Y., Zvereva P.A., Selivanovskaya S.Y. The effectiveness of co-digestion of sewage sludge and phytogenic waste. World Applied Sciences Journal. 2014; 30:1689–1693. DOI: 10.5829/idosi.wasj.2014.30.11.14234

[20] Langer S.G., Ahmed S., Einfalt D., Bengelsdorf F.R., Kazda M. Functionally redundant but dissimilar microbial communities within biogas reactors treating maize silage in co-fermentation with sugar beet silage. Microbial Biotechnology. 2015; 8:828–8:836. DOI: 10.1111/1751-7915.12308

[21] Popescu C., Jurcoane S. Evaluation of biogas potential of some organic substrates from agriculture and food industry and co-digestion in large scale biogas plant. Romanian Biotechnological Letters 2015; 20:10648–10655.

[22] APHA (American Public Health Association). Standard methods for the examination of water and wastewater. 18th edition. Washington, DC: American Public Health Association; 1992.

[23] Kapp H. Schlammfaulung mit hohem Feststoffgehalt. Munchen: Oldenbourg Verlag; 1984.

[24] European Biogas Association [Internet]. Available from: http://european-biogas.eu/wp-content/uploads/2016/01/Graph-3-Evolution-biogas.png [Accessed: 2016-04-04]

[25] Offer N.W., Marsden M., Phipps R.H. Effect of oil supplementation of a diet containing a high concentration of starch on levels of trans fatty acids and conjugated linoleic acids in bovine milk. Animal Science. 2001; 73:533–540.

[26] McEniry J., O'Kiely P., Clipson N.J.W., Forristal P.D., Doyle E.M. The microbiological and chemical composition of baled and precision-chop silages on a sample of farms in County Meath. Irish Journal of Agricultural and Food Research. 2006; 45:73–83, DOI: 10.1111/j.1365-2672.2009.04557.x

[27] Danner H., Holzer M., Mayrhuber E., Braun R. Acetic acid increases stability of silage under aerobic conditions. Applied and Environmental Microbiology. 2003; 69:562–567. DOI: 10.1128/AEM.69.1.562-567.2003

[28] Steidlová Š., Kalač P. The effects of using lactic acid bacteria inoculants in maize silage on the formation of biogenic amines. Archives of Animal Nutrition. 2003; 57:359–368. DOI: 10.1080/00039420310001607716,

[29] De Lemos Chernicharo C.A. Anaerobic reactors (Volume four of biological wastewater treatment series). IWA Publishing, London; 2007.

[30] Straka F., Jenicek P., Zabranska J., Dohanyos M., Kucnarova M. Anaerobic fermentation of biomass and wastes with respects to sulfur and nitrogen contents in treated materials. In: Proceedings Sardinia 2007, Eleventh International Waste Management and Landfill Symposium S. Margherita di Pula, 1–5 October; Cagliari, Italy. 2007. p. 1–9

[31] Hutňan M., Kolesárová N., Bodík I. Anaerobic digestion of crude glycerol as sole substrate in mixed reactor. Environmental Technology. 2013; 34:2179–2187. DOI: 10.1080/09593330.2013.804581

[32] Amon Th., Amon B., Kryvoruchko V., Bodiroza V., Pötsch E. Zollitsch W. Optimising methane yield from anaerobic digestion of manure: effects of diary system and glycerine supplementation. International Congress Series. 2006; 1293:217–220. DOI: 10.1016/j.ics.2006.03.007

[33] Leible L., Kälber S., Kappler G., Oechsner H., Mönch-Tegeder M. Biogas aus Landschaftspflegegras Möglichkeiten und Grenzen. Karlsruhe: KIT Scientific Publishing, 2015.

[34] Mahmood A., Ullah H., Ijaz M., Javaid M.M., Shahzad A.N., Honermeier B.:Evaluation of sorghum hybrids for biomass and biogas production. Australian Journal of Crop Science. 2013; 7:1456–1462.

[35] Kreuger E. The Potential of Industrial Hemp (Cannabis sativa L.) for Biogas Production. PhD Thesis, Faculty of Engineering at Lund University, Sweden; 2012.

[36] Kiesel A., Lewandowski I. Miscanthus as biogas substrate—Cutting tolerance and potential for anaerobic digestion. Gobal Change Biology Bioenergy. DOI: 10.1111/gcbb.12330

[37] Weiland P. Biomass digestion in agriculture: a successful pathway for the energy production and waste treatment in Germany. Engineering in Life Sciences. 2006; 6:302–309. DOI: 10.1002/elsc.200620128